Teaching Science to Children

Lazer Goldberg

With a New Preface by
Carollyn S. Oglesby

DOVER PUBLICATIONS, INC.
Mineola, New York

To Mari,
and to Sarah, Jonathan, and Jeremy

Copyright

Copyright © 1970 by Lazer Goldberg.
New Preface Copyright © 1997 by Dover Publications, Inc.
All rights reserved under Pan American and International Copyright Conventions.

Published in Canada by General Publishing Company, Ltd., 30 Lesmill Road, Don Mills, Toronto, Ontario.
Published in the United Kingdom by Constable and Company, Ltd., 3 The Lanchesters, 162–164 Fulham Palace Road, London W6 9ER.

Bibliographical Note

This Dover edition, first published in 1997, in an unabridged republication of the work first published by Charles Scribner's Sons, New York, in 1970 under the title *Children and Science*. A new Preface by Carollyn S. Oglesby has been added to the Dover edition.

Library of Congress Cataloging-in-Publication Data

Goldberg, Lazer.
    Teaching science to children / Lazer Goldberg ; with a new preface by Carollyn S. Oglesby.
       p.   cm.
    Originally published: New York : Scribner, 1970.
    Includes bibliographical references and index.
    ISBN 0-486-29600-8 (pbk.)
    1. Science—Study and teaching (Elementary)   I. Title.
LB1585.G623   1997
372.3'5'044—dc21   96-53258
                                                                                    CIP

Manufactured in the United States of America
Dover Publications, Inc., 31 East 2nd Street, Mineola, N.Y. 11501

Contents

Introduction   v

Preface to the Dover Edition   ix

1. The Climate   3

THE ANTI-AUTHORITARIAN TRADITION IN SCIENCE • HOW DIFFERENCES ARE RESOLVED IN SCIENCE • SCIENTISTS ARE PRODUCTS OF THEIR CULTURES • QUESTIONS ARE ESSENTIAL • A CLIMATE FOR LEARNING SCIENCE • LEARNING TO OBSERVE • UNDERSTANDING IS ALWAYS LIMITED • SCHOOL IS WHERE YOU FIND IT • TOWARD A SCIENTIFICALLY LITERATE NATION

2. Play   24

PLAY IS INSTRUCTIVE • FREE EXPLORATION • USEFUL PLAY • RESPECT FOR INDIVIDUAL INTEREST • CHILDREN CHOOSE THEIR ACTIVITIES • THE OUTDOOR LABORATORY • PLAY IS A BEGINNING

3. Games   42

A SCIENTIFIC PUZZLE • MODELS • SCHOOL AS A LEARNING LABORATORY • THE TEACHER AS A LABORATORY DIRECTOR • EXPERTS CONTRIBUTE • THE LABORATORY TAKES TO THE ROAD

4. Questions and Problems   62

ONE QUESTION LEADS TO ANOTHER • "TO QUESTION" IS A VERB • KEEPING THE SPIRIT ALIVE

5. Head and Hand   75

HUMAN UNIQUENESS • THEORY AND PRACTICE • EXPERIMENTS AND EXPERIENCE • THE GIANT PROJECT • ACTIVITY IS CENTRAL

6. Head and Heart   92

THE SEPARATION OF INSEPARABLES • CHILDREN FEEL AND KNOW

7. Error and Failure   105

INSUFFICIENT INFORMATION • INTELLECTUAL INERTIA • SNOBBERY, SENILITY, AND CERTAINTY • MAKING MISTAKES • USING DISAGREEMENTS • KNOWING WORDS IS NOT ENOUGH

8. What Have They Learned?   120

FINDING EVIDENCE OF LEARNING • TESTS PENALIZE THE THINKER • TEACHING—NOT TESTING

Selected Bibliography   133

Index   141

Acknowledgments   147

Introduction

Science for children is a fairly recent concern. In 1897 John Dewey, the American philosopher whose thinking has profoundly influenced our educational practice, argued in *My Pedagogic Creed* that science is particularly suitable in the education of children. In *Democracy and Education* (1916) and other writing, he proposed the view that science education is properly concerned with process and inquiry. In 1927 Gerald S. Craig, the pioneering science educator at Teachers College, Columbia University, published *Certain Techniques Used in Developing a Course of Study in Science for the Horace Mann Elementary School.* In this seminal work, Craig made a persuasive case that science education encompasses far more than nature study, and that it should have a more central place in children's education. In 1931 New York State produced the first course of study in elementary science education; it was the first state to do so. The following year the first series of elementary science textbooks was published. Only about one generation has passed since science for children attracted serious attention.

That attention has grown more intense in recent years. With the establishment of the National Science Foundation in 1950, a new level of recognition was given to science in the affairs of the nation. The launching of Sputnik I by the USSR on October 4, 1957, stimulated reviews of American efforts to educate children in science. By 1968 there were some twenty projects in the United States to develop programs in elementary science.

What these programs are trying to do, each in its own way, is to make Americans scientifically literate. But there is wide disagreement about what constitutes literacy, about the best means to achieve it, and about the needs and capacities of children. The only point of general agreement is that one does not become scientifically literate by committing a long list of facts to memory.

It is difficult to agree about how to educate children in science because the proposed solutions will vary according to individual concepts of the troubles and the promises of our contemporary world and according to particular visions of a desirable world and how such a world may be fashioned. Descriptions of the world in which the children are growing up today, and those of the world in which the children will be adults making their own decisions, ultimately rest upon values. Planning science programs for children without first agreeing upon their value-laden underpinnings is like trying to have a happy family outing without first agreeing upon a destination.

The realization that one must consider the characteristics of the good person, the good life, and the good world, along with the qualities of good science for children, has come late, but it has come. The inclusion of seven "values" which characterize work in science in "Education and the Spirit of Science," a statement issued by the Educational Policies Commission of the National Educational Association in 1966, was a recognition of this necessity. Although that statement does not make clear what values children are to acquire, it nevertheless indicates that science is no longer indifferent to virtue. Further recognition was given to a more humane view of science by the publication in 1969 of *Behavioral Objectives in the Affective Domain* by the National Science Teachers Association. Here the authors, Albert F. Eiss and Mary Blatt Harbeck, deal explicitly with the subject of values, attitudes, and interests, but it was not their purpose to examine particular values, attitudes, and interests beyond employing them as examples of the affective domain, and establishing their relevance to science education.

Such an examination must be made if the connection between science and children is to be both natural and humane. Man invented and developed science in his efforts to know, to savor, to explain, and to control his natural environment. From the moment of their birth, children's efforts are directed toward increasing mastery in doing these very things. Children are the inheritors of a long tradition which was acquired by mankind through the trials,

Introduction

the errors and failures, the visions and occasional successes, of men and women in all parts of the earth. What shall be distilled from this vast human experience? For clearly, children cannot relive it. What considerations shall govern the selections? If the primary obligation of educators is to fill the vacant places in laboratories, then those selections ought to be made that will equip children to fill them. But if the primary obligation is to assist children in their fumbling attempts to cultivate their humanity; if in their growth, as in great art, we sense the tremors of the future, then educators should make quite different selections. Some children may eventually choose the laboratory and a life of science, but that should be entirely fortuitous. Such choices should be the consequences of particular talents and interests, not of society's anxiety to produce cogs and wheels for a huge, impersonal machine.

The selections from science in this book are respectful of its integrity, not because science is holy writ, but because without such respect it is not possible to develop honest argument. The selections are also respectful of children, of their vast, untapped capacity to learn, of their need to find learning significant and interesting, and of their right to change the world in accordance with their own vision.

Preface to the Dover Edition

Lazer Goldberg's *Children and Science* is a book to read slowly and to relish. It was written by a man of piercing insight and gentle means, who truly understood both the nature of science and the nature of children, and how the two of them fit hand-in-glove. In reading it one becomes imbued with the idea that elementary school children can carry into later childhood the inquisitiveness and effectiveness of the preschool years, without loss of momentum and curiosity. It is a book for people who understand and appreciate the critical importance of the child's own means of constructing and organizing knowledge. In these pages one will find encouragement and direction for creating the environment that enables children's sustained self-motivation to play the central role in learning.

Children and Science is unique in its undercurrent of trust in children's willingness to learn in order to understand and grow in the world they've inherited. It stands apart in its attitude of respect for the intelligence of all ages, but especially for the intelligence of children, for their ways of learning and their boundless potential. It examines the nature of scientific discovery, the role of science in society and the relationship of educators to students. Finally, it is written not only with exceptional clarity, but with courtesy, to which we are all responsive, regardless of age.

Science is becoming more relevant to life each day, and there is no end to that in sight. Technology is now irrevocably interwoven with economy and society, making scientific literacy a prerequisite for both public and private decision-making. It is no surprise, therefore, that science education for children is a topic of intense interest and dynamic effort. Science education initiatives are flowing from teaching colleges and universities, private corporations, public service organizations and non-profit sectors. Classroom science is being reinvigorated in the light of developmental psychology. Liaisons between expert scientists and teachers are becoming more common. Programs providing teachers with summer research experience are flourishing, and science play programs are becoming available for elementary students. Science television programs are accessible and science videos are available. All of these initiatives are worthy of support, insofar as they differ from the lecturer/listener mode of the past and seek to actively engage students in awareness of science as a fact of everyday life. The objective of the best of these programs is the demystification of science.

If one defines science as the pursuit of knowledge of the natural world for the purpose of prediction and control, then children are among the most successful scientists from the day of birth through early childhood. Human infants' early efforts to organize sensations into meaningful information, acquire motor control and then native language capability are highly successful—and almost as devoid of failure as they are of teaching. Their success results from inquisitiveness and persistence, from the ability to distinguish what works from what doesn't, from the ability to relate new information to known information, and from the courage to get right into the thick of things—even to buck the system for the sake of new knowledge. These are the most basic characteristics of the scientist and this is what children are born with. Both the scientist and the child select which bits of knowledge, out of the infinite array available at any given time, are relevant and can be meaningfully incorporated into the already existing body of personal knowledge. As effective as this learning is, however, it is supplanted by almost all schools, where the child's own initiative and methods are exchanged for those of adults. As the power is lost to perform self-devised experiments and self-select important data, as book learning and seat work take the place of concrete stimuli and free exploration, as outer authority overshadows probing inquisitiveness, some of the light goes out of the eyes. In the years that follow, many children exchange their exploratory inquisitiveness for approval, attention and acceptance-related activities.

It's true that the modern school environment, especially with today's nationally standardized objective-based education pressure, is not easily adapted to the freedom of movement and choice that comes with active exploration. Models do exist for transforming science education, however. One example is the Hands On Science Outreach Program[1], which provides nonjudgmental after school science play experiences for early elementary-aged children. The educators, lay people and scientists who run the classes are trained in encouraging

Preface to the Dover Edition

children's participation in asking and answering questions during exploration. Individually and as a group, the children discover and examine data, discuss and reform ideas and eventually, reach consensus. Bringing the same spirit of shared discovery into the classroom might just be half the battle. But the ease of transforming the classroom is not the issue. The rewards of adopting this more trusting and respectful attitude toward children are real and immense, and have even been known to spark the general reform that is desperately needed in many sectors of public education, as shown, for example, by James Comer[2] in the New Haven, Connecticut public schools. In transforming the lowest-rated elementary schools to among the highest, Comer showed that content at the elementary level is much less important than *contact,* and that contact with adults who are themselves curious, enthusiastic and supportive of learning is critical. It is less important to stick information onto children than it is to consult with them in a way that allows them to remain actively engaged in collecting and evaluating information that is important to them. Education at its best is a community affair, a group association of actively learning people, where all are enriched and encouraged by peers and role models who love learning themselves and who relish independent thought. Successful creation of such learning environments may go far in bringing us to the day when children are free to learn, and can be seen and appreciated more for themselves than as credits or detriments to their parents and their schools.

In this book are all the elements necessary to keep the honesty, energy and curiosity of early childhood alive throughout elementary school age years, whether those years are spent in public, private, alternative or home school. There is respect for inquisitiveness, and the sensitivity to look at each problem afresh from each child's perspective. There is the all-important observation that mistakes are opportunities for learning that can often lead to fortuitous discovery, and that grades and tests are generally counterproductive to long-term learning. Alternative means of evaluation of progress are provided. There is honor for defensible dissent and independence; asking the question that defies a quick answer is regarded as an opportunity for even the leader to explore new territory. There is recognition of the intimate relationship between play and creativity. There is kindness

Teaching Science to Children

and encouragement for diverse learning styles, combined with a keen eye for recognition of those children for whom scientific rigor will be meaningful. A great part of the utility of this book will be to renew and reinforce the fact that, especially on the elementary level, science is terrific fun for all who undertake it in the spirit of adventure.

When Lazer Goldberg died in 1992 we lost a great and gentle leader. It is my most sincere hope that something of Lazer Goldberg's spirit and knowledge—his love of science and children—will live on and be transmitted to each of you in your working with the children in your lives. And may you continually learn from them as well.

—CAROLLYN S. OGLESBY

(1). Goodman, I.F., and Rylander, K., "An Evaluation of Children's Participation in the Hands On Science Outreach Program," Sierra Research Associates, 26 Lee Street, Suite I, Cambridge, MA 02139, May, 1993.
(2). Schorr, L.B., *Within Our Reach, Breaking the Cycle of Disadvantage,* Anchor, 1989, or Comer, J.P., *School Power: Implications of an Intervention Project,* Free Pt., 1993.

Teaching Science to Children

1. The Climate

> Slow Miss Mandy,
>   Her babies weren't fat,
> But they always wanted
>   What they couldn't get at;
> On the very top shelf
>   She put the cream in a crock,
> And she left the ladder handy
>   And the key in the lock.
>
> Hughes Mearns, "Slow Miss Mandy"

One day an eight-year-old boy observed the installation of acoustic tiles in the school library ceiling. He asked the workman what he was doing. The workman replied that he was putting in tiles so that the library would be quieter. The boy asked what the tiles did, and the workman answered that they trapped noise. The boy then came to see me.

"The man working in the library said that the tiles he's putting on the ceiling trap noise. Is that true?"

"Yes, it is," I said.

"Then, suppose I get enough of them to make a box. And inside the box I put an electric bell. And I make a small hole to pull the wire out. And then I connect the wire to a battery. And I plug the hole around the wire real tight. The bell would ring inside the box, but we wouldn't hear it, would we?"

For the first time he looked up at me. He had been busy building the imaginary box, and it had required all his attention.

Teaching Science to Children

"We probably would hear it, but faintly," I suggested, wondering what he was driving at.

"Then, suppose I let it keep ringing for a long, long time—maybe until the battery was all used up. Not much ringing would get out of the box, right?"

"No, not much," I agreed, more puzzled than ever.

"Then," he concluded with a triumphant smile, "then, when I opened the box, all the saved up sound would come out with a terrible noise."

"No, it wouldn't," I said.

"Where did the sound go?" he wanted to know.

It was a stunning question. All that sound cannot just disappear. It must register its presence somewhere. I was tempted to give an impromptu lecture on energy conversions and the conservation laws, but I resisted the temptation.

"When children practice on their instruments in the music room, can you hear them all over the school?" I asked.

"No." He shook his head.

"Why not?"

"I guess the music gets used up," he replied, raising his shoulders.

"I guess the ringing of the bell also gets used up," I said.

The boy clearly was not satisfied. I showed him a little booklet on sound. He thumbed through it and took it with him. If the puzzle he had discovered proved to be sufficiently compelling, he would find something to do about it. Perhaps he might even build the box and check the results for himself. I knew he would pursue the problem to the extent of his interest and ability. The school would provide encouragement, time, and materials.

What are the conditions of learning that will encourage children to observe the events in their common experience and to note the uncommon, puzzling qualities about them? What can adults do, or refrain from doing, to cultivate children's devotion to questions, to criticism, and to doubt? How can adults discourage the passive and intellectually opportunistic game in which the child guesses what

The Climate

answers are wanted by adults and somehow provides them? In short, what kind of learning climate will best launch children upon their science education? The history of science can give considerable direction to the search for the answers to these questions.

THE ANTI-AUTHORITARIAN TRADITION IN SCIENCE

There was a time when descriptions of the universe were adjudged valid or invalid in accordance with the status, prestige, and power of the person who offered them. When, in 1600, Giordano Bruno, the Italian philosopher, refused to accept these as appropriate criteria for truth, he was burned, the first martyr for modern science.

The scientific revolution introduced the novel idea that in matters of truthfulness about our world, social position is irrelevant. What matters is whether a description, when matched against the real world, fits properly. If a person makes statements that consistently fit well, they make him an authority. To insist that one ought to be believed because of name, fame, caste, or station merely makes one an authoritarian. Science respects authority, but it is fundamentally anti-authoritarian. (The anti-authoritarian position of science has a corollary. If social status is irrelevant in establishing the validity of assertions about the natural environment and about man, then any person who learns to do scientific work may make a contribution.)

The historical forces that came to life with the death of the feudal order in the sixteenth century gave birth to this new scientific spirit, but it was not an easy birth, not without pain and difficulty. Nicolaus Copernicus, the Polish astronomer, lay dying in 1543 when he saw the publication of his ingenious work, *De revolutionibus orbium coelestium libri VI* (Six Books Concerning the Revolutions of the Heavenly Spheres). This work explains the motions of the heavenly bodies with far fewer assumptions than the

5

Greek astronomer of the second century, Ptolemy, found necessary. Copernicus imagined the motions of the solar system from the vantage point of the sun. It was an elegant idea, but it was also a dangerous one. Georges Joachim, Copernicus' twenty-five-year-old Austrian-born disciple (who used the name Rheticus), had convinced his master to publish. The abandonment of a geocentric universe was attacked from all the high places. What is revealing is that none of these attacks accused Copernicus of presenting a model that conflicted with astronomical observations, rather the conflict was with the philosophical and theological preferences of those in positions of power. Galileo Galilei, the Italian astronomer and physicist whose work at the end of the sixteenth and beginning of the seventeenth century helped to lay the foundations of modern science, confirmed the Copernican theory by looking at the sky with a telescope of his own design—and was imprisoned for his contribution. In *The People, Yes* (1936), Carl Sandburg conveys the difficulties men endured in the struggle against authoritarianism:

> In the folded and quiet yesterdays
> Put down in the book of the past
> Is a scrawl of scrawny thumbs
> And a smudge of clutching fingers
>
> And a thinker locked into stone and iron
> For saying, "The earth moves,"
> And the pity of men learning by shocks,
> By pain and practice,
> By plunges and struggles in a bitter pool.
>
> In the folded and quiet yesterdays
> How many times has it happened?

Authoritarian decree may have caused voices to be stilled and books to be banned, but it was no good at all in building engines,

The Climate

combating disease, or in increasing the yield of agriculture. The new ways of science proved exceedingly effective in providing the understanding that made it possible to do these things on a scale undreamed of before. The introduction of modern science could not eliminate self-interest. Nor did it eliminate disagreement. On the contrary, it encouraged it. But the cost of discovering a new orderliness, of perceiving the novel, of honesty and daring, was no longer so dear. And now the disagreements, whether they result from differences in point of view or from self-interest, are subject to test by the members of the scientific community.

HOW DIFFERENCES ARE RESOLVED IN SCIENCE

Sir Isaac Newton, the great English scientist and mathematician who was born in 1642, the year that Galileo died, believed that light was a stream of particles because it gave sharp shadows and traveled in a straight line. Christian Huygens, a Dutch physicist and astronomer and Newton's contemporary, thought that light consisted of waves. Newton's scientific authority was so great that his particle model of light went unchallenged for a hundred years. Experiments early in the nineteenth century by the English physicist and physician Thomas Young, and by the French physicist Augustin Jean Fresnel, corrected Huygens' view of light as longitudinal compression waves, like sound waves, and suggested that they were transverse, like water waves, but they supported his wave model of light. This model fitted very well into the electromagnetic theory of light developed later by the Scottish mathematician and physicist James Clerk Maxwell. However, in this century, the concept of the quantum developed by the German physicist Max Planck, the explanation of the photoelectric effect by the German-Swiss-American physicist Albert Einstein, and experiments with X rays by the American physicist Arthur Compton which showed that when these are passed through materials they behave like billiard balls, gave a new turn to the old disagreement.

Teaching Science to Children

It turns out that light and matter have both particle and wave properties. The argument has by no means been resolved to everyone's satisfaction. New views are invited. Their validity will be tested by practice.

This brief account of the disagreement about the nature of light suggests what is characteristic of the manner in which differences are resolved in all the sciences. In geology, the argument during the early part of the nineteenth century was between neptunists and vulcanists. The leading neptunists, Abraham Gottlob Werner, a German, and his Scottish pupil Robert Jameson, held that the solids on the surface of the earth were precipitated while the water which had covered the entire globe gradually disappeared. Two other Scottish scientists, James Hutton, a geologist, and John Playfair, a mathematician and geologist, insisted upon the importance of heat in explaining the hardening of sediments, in surface eruptions, and in the underground formation of granite. Werner was certain that his point of view was correct and refused to hear of any evidence that contradicted his theories. Nevertheless the evidence accumulated. William Smith, an English geologist and canal-building engineer, traced fossils and their beds. Sir Charles Lyell, still another Scottish geologist, found evidence of processes which are still operative in rocks dating from ancient times. Werner's view simply did not stand up to an examination of the earth itself.

Eventually differences in science are resolved, but agreement does not come easily. In 1628 William Harvey, an English physician, published *On the Motions of the Heart and Blood,* a little book that laid the basis for modern physiology. Forty years later Francesco Redi, an Italian physician and poet, read the book and was inspired by Harvey's suggestion that living things grew from eggs or seeds too small to see. Redi performed a lovely experiment, one of the earliest instances of the use of excellent controls in biological research, and concluded that maggots are not created by spontaneous generation, but by flies which lay eggs. In 1748

The Climate

the English priest-naturalist John Turberville Needham performed his own experiment in which he brought mutton broth to a boil and then placed it in a corked test tube. After a few days he found microorganisms in the broth. He believed that he had proved the case for spontaneous generation, at least for microorganisms. The Italian biologist Lazzaro Spallanzani believed that Needham had not really sterilized the broth, so twenty years after Needham's experiment he performed one of his own. No microorganisms appeared. It would seem that this might end the argument, but for another century the question remained in doubt. It was not until 1860 that the French chemist Louis Pasteur provided brilliant experimental evidence that finally put the old argument of spontaneous generation to rest.

Old theories die hard. Unlike so many old people, they are almost never left alone and neglected, but are protected and defended to the very last. One result is that new ideas in science frequently have very difficult births.

SCIENTISTS ARE PRODUCTS OF THEIR CULTURE

Scientists are not selfless saints. They compete for priority in discovery and for recognition by respected colleagues. They have accused each other of unfair practice, and they have questioned each other's motives and honor.

There is no reason for surprise that some of the greatest figures in the history of science should display such behavior. It would be surprising were it otherwise. It is to be expected that cultures whose aspirations to individual salvation, as expressed in the old Salvation Army song, are "The bells of Hell go ting-a-ling for you, but not for me," will affect even the best minds to look to themselves. It is not startling that Galileo, Huygens, and Newton, giants in science, were infected by the acquisitive and competitive drive of the cultures that reared them.

Teaching Science to Children

The efforts of scientists to win, to have their work acknowledged, and to achieve fame, simply mean that like their fellow citizens, they are men and women of their time. They come in all the usual sizes, both physically and ethically. Their needs and their wants are not unlike those of the rest of us. Men scientists admire pretty women, and sometimes succeed in marrying them. Like other parents, they enjoy playing with their children. Like other professionals, most scientists have many interests besides their scientific work. They are no more or less sociable than others. If some scientists are prominent in the political news of the day, it is because their special knowledge is relevant to the principal issues of our time.

Whatever may be the characteristics that mark the uniqueness of an individual, if he joins the community of scientific workers, he learns to share with them a common way of regarding the world, a style for searching among the countless disparate details by which we are surrounded to find unity and meaning. He learns that there is no paradigm for the solution of problems, no neat series of steps that will lead him from question to answer. He discovers that the activity of science is characterized by neither a method nor by pot luck, but by the indissoluble marriage of theory with practice. If he is imaginative and his mind is prepared, he may find explanations, regularities suggested by practice, and these insights may illuminate new practice.

The history of science and the biographies of scientists are most persuasive that scientific activity does not exclude any willing participant. There is a place for people with diverse personalities, temperaments, and dispositions. They may come from all segments of society, derive from all ethnic origins, and be committed to conflicting ideologies.

Joseph Priestley, the discoverer of oxygen, was poor, suffered from a serious speech defect, was a radical political dissident who supported the French Revolution, and was compelled to flee from the bigotry of his native England to live out his last years in the

The Climate

United States. Antoine Laurent Lavoisier, who laid the foundations for modern chemistry, was well-to-do, a brilliant conversationalist, a vain man who failed to give Priestley credit for his work on oxygen. He died on the guillotine.

Joseph Henry and Josiah Willard Gibbs were two great American scientists of the nineteenth century. Henry, who discovered how to convert magnetism to electricity, was outgoing and romantic and might have been an actor instead of a scientist. Gibbs, who constructed the mathematical foundations of chemical thermodynamics, was quiet and withdrawn.

Toward the end of the nineteenth century the German-American physicist Albert Abraham Michelson designed beautiful apparatus to study the postulated ether by measuring the anticipated changes in the velocity of light as the earth made its annual trip around the sun. In 1905 Albert Einstein, whose principal tools were pencil and paper, or chalk and a blackboard, announced his Special Theory of Relativity, which assumed the constant velocity of light in a vacuum and eliminated any need for an ether.

Benjamin Banneker, a mathematician; Ernest Everett Just, a zoologist; William A. Hinton, an entomologist; the botanist George Washington Carver; Charles R. Drew, a surgeon; and Percy L. Julian, a chemist, all contributed significantly to the development of American science. They were Negroes, as were the technologists Jan Matzeliger, Norbert Rillieux, and Granville T. Woods.

The record speaks plainly. The ability to deal imaginatively and creatively with change, that is, to engage in scientific activity, is latent in many kinds of people. Of course, few will make great discoveries, just as few will write great novels, paint great paintings, or compose great symphonies. Little is presently known about giftedness, but it does appear that people are selectively endowed with it. That need not prevent anyone from participating in various activities, including science, enjoying them, and gaining sufficient competence to appreciate and criticize the work of professionals.

Teaching Science to Children

QUESTIONS ARE ESSENTIAL

To participate in scientific activity one need not be a walking encyclopedia—one need only be alert to possibilities for investigation. There are any number of questions that can be asked about events in common experience. The sad fact is that for most people they remain unasked.

Someone whose hands are cold will sometimes cup them and blow into them for temporary relief. When the same person carelessly touches a hot pot, he will blow on the painful place while he rushes to the cold water. The same breath blows warm and cool air. Is it not very strange? Frequently people suggest that the temperature of the breath does not change, but only appears to do so when compared with the temperature of the skin. The temperature of the breath, they say, remains constant. Why then, do they shape their mouths differently to blow warm and cool air?

There is no more popular toy than a ball. Everyone has played with one at some time or another. Why does it bounce? And why do glass marbles, steel ball bearings, plastic or ivory billiard balls, also bounce? What do all these have in common? Why do they behave differently from wet clay, which lands with a thud and just lies where it has fallen?

People do not usually discern the puzzling features in common experience. There are many contributing factors for lack of sensitivity to the environment—hurried lives, a variety of stimuli competing for attention, the order of priorities that defines successful living, and other sufficient reasons. Schooling has not helped in overcoming these barriers. Rather, it has contributed yet another barrier. How are adults to find interesting questions when they were raised with the conviction that questions are posed by other people, and that the answers to them, should one care to know, are neatly accumulated in textbooks? The schooling that should have made them feel more at home in their world has served to make them strangers. It is not a comfortable feeling to be a stranger.

A nine-year-old boy made the happy discovery that carrots do

The Climate

not arrive in the world devoid of greens, neatly bundled in plastic bags. He was happy because a tiny bit of his world was less strange to him, and he is not really very different from the rest of us. Look at the surroundings within your home. Think of the electric power, the water, the food, the furnishings, the clothing, the kitchen utensils, and the very structure of the house. How familiar are they? Not their superficial appearance, but their sources, the mechanisms that make them useful, their construction, their history. Is it really enough to satisfy a human being to know that switches are turned this way and that? Have Americans been so altered by technology that conveyor-belt existence is to be their natural form of living? Every newborn infant attests that this is not so. With relentless searchings and probings he tries to understand this world into which he has been born. He continues to search until he is stopped. At some point about the age of eight or nine, he finds that his boundless need to try, to test, to question, to feel, to know, gets him into difficulty, makes his teachers and his parents unhappy, and therefore makes him unhappy.

The child becomes passive and unhappy because isolating problems, finding interesting questions and puzzles, and pursuing these the best way he can, activities that are natural to growing children, are not the usual preoccupations of his schooling. More commonly, schooling means the pursuit of neat, discrete answers to questions someone else has already posed. Such a system allows but little room for disagreement and still less tolerance for error and failure. Not to know is a confession of inadequacy—it means that one has not listened carefully, has not read the book, has not done the assigned homework, has a bad memory, is stupid, or at least indifferent. In this climate it is humiliating and fearful not to know. Yet, there are many times when a child ought to be able to say with dignity, clear and loud, that given the conditions, or with the available information, he does not know.

Consider a typical example. Is there a child who, somewhere between the ages of seven and ten, has not "learned" about day and night? The model is a familiar one, a stationary sun and a rotating earth. This model conflicts with the real world that is familiar to children. It is a model even sophisticated adults found it difficult to accept for a long time. Yet, children are expected to deny their very real experience for a model that scarcely makes sense to them. Why should they believe it? What arguments have been produced to support it? Teachers have learned, especially in scientific matters, not to doubt, not to trust their independent intelligence, and now it is the children's turn to learn that lesson. Adults, particularly their books, are not to be doubted.

The rare child who feels free to insist that the model is incorrect because he has seen that it is the sun that travels through the sky and the earth that is stationary is thinking clearly, reporting accurately, is honest and daring. (Such a child is usually labeled "troublemaker.") But if no convincing evidence is offered to him, why should he sacrifice his own conclusions? He is in the best tradition of modern science. Those who insist upon a response no more rational than that "it says so in the book" are laboring in an authoritarian world that preceded the scientific revolution.

When adults, including teachers, are asked to justify the Copernican model, few volunteer. Was the lesson in any sense a contribution to their scientific education?

A CLIMATE FOR LEARNING SCIENCE

In order to invite children to engage in science activities and to encourage them in their efforts, it is critical that an appropriate climate be created. The essential features of such a climate, derived from the practice of science and the experience of children, are anti-authoritarianism and democracy; high tolerance for dissent, argument, error and failure; regard for esthetic reward; absence of fear and humiliating measurements; emphasis on cooperation rather than competition; respect for manual as well as intellectual effort; and, above all, interesting and significant activity. It is not a climate in which "anything goes." Rather it is one

The Climate

which is humane and reasonable. It is a climate in which children cultivate valid criteria for choosing among alternative beliefs.

Children are not shackled, but are encouraged to use their gifts to the fullest, and opportunities are provided for them to do so. There is more than one way to put fetters on a child's spirit. A common practice, one that goes almost unobserved, is to place time above task in the order of priority. The arbiter of movement from task to task is the clock or the bell, not the inherent requirements of the activity. A child may be engrossed in his most serious work; no matter, he must leave it and go on to something else that has been scheduled for this time. Where the bells are not electric ones, they are internal, psychological ones that ring inside the teacher. When these bells sound, she claps her hands and announces that the present work must be stopped, put away, and the next work begun.

Later, after the child has learned that no task is so vital that it cannot be deserted, his parents wonder why so many "unfinished symphonies" clutter the house. Still later, when the children have grown up, writers debate the causes of lack of commitment among the American people. It would be naïve to suggest that the school is solely responsible. It would be equally naïve to suppose that the school does not contribute to an attitude that no task can be so absorbing as to have earned a person's deepest dedication.

There are unavoidable signals that interrupt work. At school there are fire bells and a time for dismissal at the end of the day. At home there are mealtimes and bedtimes. Children can understand the need to comply in matters of this kind, and they will plan their work so that it suffers the least interruption. They can also learn to respect the adult's authority in matters of safety. For example, children consider it reasonable that they may not experiment with household current without adult supervision. They also agree with the provision that one person's activity may not interfere with that of another because they resent interference in their own work. Children understand necessary social incursions upon their autonomy; they are not irritated or frustrated by them as they are by the restrictions the clock places upon the emotional investment in a task.

Intellectual restrictions cannot be justified if knowledge is to be pursued honestly. No question is a bad one if the questioner really wants to know. The caveat is for those who like to ask questions in order to mark their presence. No one is beyond argument and disagreement—not the teacher, the parent, the principal, the mayor, the governor, or even the president. What is expected is that reasonable argument will be presented, supported by evidence of some sort. Such intellectual permissiveness means that no institution, however hallowed by time or reputation, is immune to criticism. Such criticism extends to the science class itself, to the school, and to the home.

When children experience such an atmosphere for the first time, they are incredulous. It is too liberating. There must be a fly in the ointment somewhere. Once they are convinced that it is not some sneaky ruse, a new trap to catch them unawares, they become intoxicated with the new freedom. The intoxication lasts a short time, but there are no ill aftereffects. The children are then ready to settle down in a workmanlike atmosphere. Thy are ready to learn.

This climate must be constant and pervasive. It cannot be opened and closed like a faucet without confusing children and putting them in doubt about the sincerity of the entire enterprise. Honesty and integrity in social interactions, intellectual vigor, and esthetic sensibility pervade the entire learning experience in any serious attempt to infuse children with a sense of wonder about their world and a desire to pursue its mysteries.

In such an atmosphere, all disciplines receive full respect. Mathematics is not drudgery and drill, but lively, imaginative, and sometimes elegant. Symbols and operations are not thrust upon children. They are derived from history and experience with materials. What is new in this "new" mathematics is not the intonation of a modern Gregorian chant that pays homage to set vocabulary, the commutative, associative, and distributive princi-

The Climate

ples, and other sacred words. What is new is the active search for the patterns that underlie these words—a search that builds solid competence and gives intellectual power and esthetic reward.

Art experience is not relegated to the periphery of learning but is made integral with it. Literature, painting, sculpture, music, dramatics, the dance, and the crafts are recognized as ways of finding out and share equal prominence with cognitive pursuits. In this climate learning is not an emotional and esthetic wasteland decorated with artistic lace curtains. It is as much preoccupied with feeling as with knowing.

LEARNING TO OBSERVE

The introduction to Harvey's little book on the circulation of the blood warns: "It were disgraceful ... did we take reports of others on trust, and go on coining crude problems out of them, and on them hanging knotty and captious disputations. Nature herself is to be addressed; and the paths she shows us are to be boldly trodden, for then, and whilst we consult our proper senses ... shall we penetrate at length into the heart of her mystery." This has not been the usual procedure in children's science education. In fact, the general practice has been to coin "crude problems" and ignore the real world. Children have been required to take too much on trust, and their attention has been drawn far too much to the teacher and the book, and too rarely to the world itself.

The proper climate for learning science is invigorating, vital, and affirmative; in it a child learns to thrive on differences, contradictions, and irregularities. Children raised in a complex technological world where so much escapes their notice must be helped to cultivate their perceptions and thereby gain a new and richer view of their environment. Too much intervenes between them and their surroundings. Few children look at the sky, at the drops of water coming from the faucet, at the house fly or the house plant. They do not look at these and many other features of their

Teaching Science to Children

environment because there is no particular reason for them to do so. Nor can they know what a loss they have sustained.

In contrast are children raised in cultures where man's relationship with his environment is clear and direct. The Pygmy child walking along the forest path learns to be observant of the finest detail. Noticing is a matter of life and death for him. For example, when he steps over a log, he must be able to notice the presence of a snake and freeze in position. The success of his lessons is testified by his survival. A Kalahari boy learns to recognize the signs of the giraffe's presence. Failure to learn his lessons well may one day mean hunger for himself, his fellow hunters, and for the entire community waiting for the meat. Children in these cultures learn principally by observing the adults. It appears natural that children raised in circumstances that place such a high premium on knowing the environment intimately should become keen observers, whether they live in the forest, on the prairie, or in the frozen wastelands of the Arctic. It seems equally natural that children raised in circumstances where the environment is not a matter of acute concern, or so it would seem, should be less observant. If this is accepted as inevitable, our children will remain ignorant not only of their natural environment, but of science and technology, which stand between them and the environment. But it is by no means inevitable.

A climate designed to cultivate observers should struggle against the habit of substituting names for involvement in activity. Children's vocabulary, toys, and television-watching are often cited as evidence of the generational growth in science-mindedness. Adults are impressed when children speak of rockets, orbits, moon shots, atoms, molecules, action-reaction, and other sophisticated things, but the ready use of these words can give a false impression. There is nothing scientific about mouthing words. It is much easier to talk of hovercraft or to manipulate a plastic working model than to disassemble a lock to see what makes the tumblers work, and it is less scientific. It may be *au courant* to watch space

The Climate

patrols on television, but it is less scientific than to sail a modest craft along the sidewalk after a heavy rain. In learning to watch what others have chosen to commend to their passive attention, children's senses have been dulled. When a child gives the name for something the adult feels rewarded by this evidence of his competence as a teacher. In the process the child's mind is dulled; he may well have responded with a word he does not even understand.

Those who teach science should instead aim to sharpen children's senses and minds. One way to accomplish this is to engage them in activities which they can themselves select from a variety of activities which knowledgeable adults have provided. Such activities compel children's attention and heighten their powers of observation. They simply must observe carefully if they are to rescue the baby fish, peel the layers of mica, find the star Alcor next to Mizar in the handle of the Big Dipper, or find out which half of the earthworm disappeared in the soil.

UNDERSTANDING IS ALWAYS LIMITED

There must be times when the adult, parent or teacher, says "I don't know." It is inconceivable that one can remain honest and avoid such an admission at one time or another. That is not a bad thing. It is salutary for children to learn that a parent, or even a teacher, than whom no wiser person has ever lived, does not know everything. This realization suggests to children the immensity of knowledge and the inevitable need to admit ignorance. But admissions of ignorance ought not to be necessary too frequently, especially by the teacher, because children may then lose confidence in the teacher's authority.

A teacher needs to know as much as possible, not for ready answers, but to recognize opportunities for teaching. It is not expected that everyone should be able to trace modern English

Teaching Science to Children

words to their Old English or other origins, but it is expected that those who assume the responsibility of engaging children in the learning adventure should know that the English language has a history and that its spelling has had a development which can be traced. When a child asks about the silent letters in words like "know" or "knot" and is told only that these spellings are some of the strange ways of English words, the answer implies that his question does not merit investigation, that it is beyond reason and sense. It makes little difference whether the answer is given with a smile or a frown. The effect is the same.

No adult can be expected to know the precise answers to the variety of questions that children may ask. Nor is this necessary. Teaching is not like filling empty buckets. It is directing and redirecting children's efforts so that they may learn what they need to know and find the process continually inviting.

Nevertheless the adult who assumes the responsibility for providing children with science experiences of good quality ought constantly to add to his fund of knowledge. It is certainly important to know—the more the better. But it is also indispensable to feel, to catch the flavor of scientific activity. American physicist Richard Feynman, the Nobel Laureate, once described how he first became interested in science. He did not speak of sophisticated equipment or of courses in physics, and certainly not of grades. Instead, he spoke very movingly about his father, a salesman, who used to spend his short vacations with his family in the mountains. Here the father took his son on walks and shared with the young, bright, impressionable boy the wonder he felt about natural events: how the birds succeed in coming back to precisely the same place each year, how the life cycle in the woods seemed to satisfy all the living things that depended upon it, and other remarkable puzzles. I do not know whether the father knew the names of the various birds, trees, rocks, and flowers. Richard Feynman did not mention it. Apparently, it was not important. What was important was the spirit the father transmitted. Like good

The Climate

knowledge, the feeling endured. Richard Feynman did not forget it.

SCHOOL IS WHERE YOU FIND IT

The physical plant is sometimes cited as a limitation in the development of a stimulating climate for science education. It is true that children need a headquarters, a place where they can organize their thoughts, make their plans, work with apparatus and build their models, draw their graphs, read books, listen to each other's findings, and reflect upon their experiences. But those experiences need not and should not be restricted to a room in a building. There are opportunities to observe interesting and significant phenomena wherever one happens to reside.

Every large city is good hunting ground. New York City would appear the least likely place to offer fertile terrain for field trips in natural history. Yet, in *A Natural History of New York City* (1959), John Kieran reports a great variety of rocks and minerals, 230 species of birds, and an endless array of flowers, lichen, fungi, and fern. He also records many examples of microscopic and macroscopic freshwater life, trees and shrubs, amphibians and reptiles, and mammals. And across the Hudson River, on a 75- by 100-foot lot in Ramsay, New Jersey, Frank E. Lutz collected 1402 species of insects.

In addition to the natural phenomena, the city furnishes many examples of men at work providing society's needs with the knowledge derived from science: excavations are being dug, buildings are going up, bridges are being built, aircraft take off and land, and manufactured goods of all kinds are being produced.

How few of the thousand hours children spend in school during a year are spent outside their classrooms. In one school district where the elementary schools are within walking distance of the ocean, I found only one teacher who takes his children for science investigations on the beach. Unhappily, that school district is fairly typical. What a pity. In *The Sense of Wonder* (1965), Rachel Carson suggests the possibilities for exposing the curious, searching, wondering child to the variety of experiences at the ocean shore.

TOWARD A SCIENTIFICALLY LITERATE NATION

There are no objective conditions to prevent the cultivation of a generation of "scientific minded" children. Society must want to do it. The fact that the United States is a science-powered culture will not, of itself, do it for us. Medieval Europe produced great cathedrals, but architecture was not a popular interest. How many were preoccupied with philosophical interests in classic Greece? When it is said that the Renaissance witnessed a tremendous outpouring of creative work, how many were involved in creating? The magnificent creations of a nation or a period point to the possibilities of what people are capable. The vision that produced them was a product of the time and the place that few shared. It was not possible that it should be otherwise, nor was it crucial. It is different now. Children in America must learn science because science powers their world—whether for good or for ill will depend upon decisions they will have to make. How can they make such decisions if science remains remote and incomprehensible to most of them?

America has not yet cultivated a generation which is scientifically literate. Philip Morrison, David Hawkins, and Robert Karplus, among others, have pointed the way, and David Webster has demonstrated that where the conditions are at all accommodating, where the climate is at all propitious, that way is most rewarding to children and to teachers.

There is a danger that statements about abstracts like the climate for learning, the flavor of science, or spirit and feeling will be met with suspicion or skepticism in our very pragmatic society. Though they are ambiguous, these terms are not mystical. They are

The Climate

necessarily ambiguous because, like artistic experience, they are somewhat different for each individual. They are not mystical because they do not transcend ordinary human knowledge and capability.

What is wanted is the will to organize a climate for children's science learning. In such a climate children continue their play. The games whose rules they learn elicit their most intense participation. There is no shame attached to error and failure, and fear has been cast out. Interesting errors are admired, and perceptive questions are applauded. Task governs time, and there is freedom to make, to think, to remake, to chat, and to dream a little. It is a place where dissent and independence are honored, where thought is not deprived of feeling nor art of thought. Above all, it is a place of diverse activity—social, intellectual, artistic, manual. It is a place where children transform bits of their environment and in the process transform themselves.

2. Play

> School is over,
>   Oh, what fun!
> Lessons finished,
>   Play begun.
> Who'll run fastest,
>   You or I?
> Who'll laugh loudest?
>   Let us try.
>
> Kate Greenaway, "School Is Over"

Mark Twain expressed a popular conception of play in *Tom Sawyer*. After the whitewashing episode, Twain informs the reader that had Tom "been a great and wise philosopher, like the writer of this book, he would have comprehended that Work consists of whatever a body is *obliged* to do, and that Play consists of whatever a body is not obliged to do." This distinction between work and play is not as helpful as it appears. How should we describe what infants do when they examine their hands, observe objects placed in their line of vision, stick their toes in their mouths, crawl, examine, and the thousand other things that infants do? They undoubtedly feel a compelling need to do these things. Are they working? In what sense is Nelson Rockefeller obliged to engage in politics? Is he playing? When Henry Moore first began carving holes through materials to find a new dimension for the sense of surface, was he obliged to do it? Was he working or playing? James D. Watson, the Nobel prize-winning American biologist, writes in *The Double Helix* (1968) that "All we had to do was to construct a set of molecular models and begin to play—with luck, the structure would be a helix." Is he using the word "play" carelessly, or is there a sense in which it describes accurately what they intended to do?

Is the serious amateur in sports playing or working? There are times when most adults feel a compelling need to change from their accustomed activities. The "need" for vacations does not change their playful character.

The widely held view of two distinct activities, work and play, does not take account of these questions. That view, derived from the Puritan tradition, conceives of work as a solemn obligation imposed on men for their salvation. Children, being inherently sinful, had to be made to labor in God's vineyard so that, ennobled by their own industry and virtue, they too might be saved. Activity that was not directed to eternal salvation was idling, and the best cure was a whipping. The contemporary perspective is different. What is desperately needed is human activity which will save us, if not for eternity, at least for the life of our planet.

The fact is that work and play are not clearly distinguishable activities, but different aspects of activity. *Soap Bubbles* (reissued in 1959), a delightful book composed of the record of three lectures given before a juvenile audience in 1889 by the English physicist and inventor C. V. Boys, suggests that there is a continuum between play and work. "I hope," he writes, "that none of you are yet tired of playing with bubbles, because, as I hope we shall see during the week, there is more in a common bubble than those who have only played with them generally imagine." When we consider science and children, it is the meaning of play as part of larger, more inclusive activity that is significant.

PLAY IS INSTRUCTIVE

When children are little, they learn to control their limited world by experiencing it, experimenting with it, and by planning new experiences. In the course of these activities, they accommodate largely to a reality within. External reality has little significance. It is during the period of his life when he accommodates least to the compelling factors of a very real world that the child learns at a rate never again equalled. It is then that he learns language, among the most complex of all human activities. And he learns it by playing among other human beings.

With continued play, speech becomes less egocentric; the child's inner and outer worlds become related. He learns that if he drops a toy, it can be picked up to be dropped again. Play stimulates his imagination. He tries to do absurd things, commits error after error, and except for errors that threaten his well-being or tax the patience or finances of the household, his errors are greeted with smiles, laughter, hugs and kisses. He explores every nook and cranny of his environment, delighted by novelty and expecting no rewards for his inquisitive pursuits. The stimulation of the search is its own reward. His explorations are totally free until he is admonished that this is "bad" or "dirty." By means of play and interaction with others, the child's growing perceptions bring order to his world. He gains a new sense of himself and of his powers. His reactions are plastic, ready to strike out in novel directions. Unless crushed by repressive response, he is endlessly curious.

In cultures that provide no formal schooling for their young, the plasticity of children permits them to become successful adults. They learn through play and imitation, Clyde Kluckhohn, the noted American anthropologist informs us in *Mirror for Man* (1963), to become individually "productive, socially useful, to increase the group's wealth and strength," and they learn to promote group harmony.

In many parts of the world, an incredible array of skills, some of them very complex, has been acquired without formal schooling and without professional teachers. The Jivaros of Ecuador can set broken bones in casts made of chicle. Some Navajos have preserved more than 500 paintings by memory alone. Many peoples have learned to remove the prussic acid from manioc, a relative of

Play

our tapioca. Each reindeer-herding Lapp can distinguish his own reindeer from among all the others in the land. A Pygmy child, "a little slip of a girl, or a boy too young to walk," can puff at an ember and start a blazing fire, something the American anthropologist Colin Turnbull confesses in *The Forest People* (1962) he never learned to do. The Ojibwa woman of Canada can cure meat and fish, set out nets, trap rabbits, embroider with porcupine quills, tan leather, roll twine, brew dyes, make her own utensils, and weave nets, mats, and robes.

Surely there are lessons to be learned from the vast experience of mankind in raising its young. The young of nonliterate peoples have learned, during thousands of years, by means of play, imitation, and instruction in ritual, to become that which each child would want to become could he, as an adult, see himself in retrospect. With minor exceptions, these cultures guarantee every child a place which in some essential way will weld the continuity of the culture.

America, however, is not a nonliterate culture. It is a highly technological and complex society, conditioned by its own history and its own circumstances. One of the prime functions of its schools is to produce children who are different from their parents —to break out in new directions rather than to weld continuous links, to build new worlds rather than to take models from the past. Despite the vast differences between the lives of the people of the United States and the lives of people living in nonliterate cultures, Americans should also wish each child to become productive and socially useful, to increase the group's wealth and strength, and to promote group harmony. Because America's decisions affect the lives of people around the entire globe, the promotion of group harmony here implies a concern for the entire human family.

In nonliterate cultures children learn to become competent as perceived by each culture. If they survive, children *grow* into competent adults. They are not dragged. Models for their emulation are clearly visible. The traditions for successful living are a matter of daily practice in clear view of the young.

Teaching Science to Children

In technological societies, children must grow into competence in a setting where practicing adults are hidden, where more often than not a child does not know what his father does—"He goes to the office" or to some other place. Under these circumstances, children's play is rarely associated with the activities that will engage them as adults. Shooting insects with thorn-tipped sticks is preparatory play at hunting that cultivates the future big game hunter. Playing cowboy and Indian cultivates nothing of permanent worth.

American children do have a family experience. Their games at house play prepare little girls to become women and little boys to become men. The practitioner with whom most children have direct experience is the family doctor, so they play doctor, which at least gives them the opportunity to inspect each other's bodies. Left undirected, children's play is largely dominated by the current successes on television and the toys these programs popularize. Picture a group of ten-year-old boys dressed in white sheets, heads and faces appropriately covered, rubber gloves on their hands, "operating" on a frog. The children are obviously absorbed by their roles. It is good dramatic play, a fitting satire of the television medicine show. But of course their play has very little to do with frogs—the frog is but a prop. Their play has as little to do with biology as space games have to do with celestial mechanics.

Children's play is derived from their experience and their imagination. Much of the play that children enjoy is useful, joyous, and necessary. Under almost any circumstances except where children are seriously deprived nutritionally, they organize games which test their language, muscle coordination, daring, imagination, and endurance.

FREE EXPLORATION

The natural desire of children to play can be utilized in their science education. In fact, play must be an aspect of their science

Play

learning because playfulness is a condition for understanding science. Play in this sense is serious but, I hope, not solemn business. For those to whom play sounds dangerously like idling away the time when there is nothing better to do, think about the dancer moving freely in preparation for choreographic ideas, or the painter sketching with charcoal, rubbing out, adding a line here and there, or the composer at the piano, striking a chord or two, shaking his head, striking another chord, and listening to the combined sounds of an orchestra in his mind's ear. The play that is intended to engage children in learning science is a little like that.

Play is not only for the slow or unwilling, but for the bright and eager as well. Scientists played when they were children. Newton cut sundials on stone and his name where he shouldn't have, built kites, lanterns, dolls' furniture, and collected herbs. Charles Darwin, the English naturalist, liked to make up stories that he could produce variations in crocuses, polyanthuses, and primroses by watering them with colored liquids. Darwin's contemporary, Dmitri Ivanovich Mendeléev, the Russian chemist who devised the periodic table of the chemical elements a century ago, collected stones, flowers, and insects when he was a boy. Konstantin Eduardovich Tsiolkovsky, the Russian physicist who was the first to suggest the idea of a space station early in this century, and on whose tombstone is written, "Mankind will not remain tied to earth forever," loved to indulge in daydreams when he was little and even paid his younger brother for listening to them. The number of scientists who were collectors when they were children is impressive. Collecting for children is a way of playing with the world. It is best done outdoors, away from school. But there is also a playfulness that belongs in school.

USEFUL PLAY

Adults must consider carefully the things they offer to children for play if they are to contribute to science learning. The materials for play must meet the same test that J.D. Bernal, the English crystallographer, sets for scientific problems designed to give experience in the processes of science. The materials must be interesting to the child, they should have significant ideas built into them and foster real, that is, the child's own, solutions.

Some materials and the activities they encourage embody all the necessary conditions for most young children. Building with wooden blocks; water play with a large basin and containers of various sizes, clear tubing, plastic syringes; baking and cooking; caring for guinea pigs, rabbits, frogs, toads, salamanders; planting from seeds. These are all interesting to children from the nursery years through the early grades. Block play is discarded earlier than it would be if we let the children decide. Some children are interested in caring for animals all through the elementary grades and later, if they are allowed to. Many of the environment's significant features are embodied in experiences like these to which children can assign their own meanings. "We baked bread and the oven made it hard, but we cooked applesauce and the apples got soft." "The guinea pig must like me because he holds tight when I put him against me, but I guess he likes them too, because he holds tight when they pick him up." The experiences are real. And whether the children are already reading or not, they ought to be surrounded by beautiful books related to these experiences.

All this is nothing new to good nursery schools, nor to teachers who by temperament and training have been able to enter the world of early childhood, there to design environments in which children may flourish. Whether their educational and philosophical mentors are Friedrich Froebel, Maria Montessori, Susan Isaacs, Arnold Gesell, John Dewey, A.S. Makarenko, Jean Piaget, or A.S. Neill, good teachers of nursery-school children make provisions for growth, not for a tug-of-war. To see the joyous, imaginative, persistent capacity of children for learning, get thee to a nursery.

Maintaining children's spontaneous and enthusiastic involve-

Play

ment in learning grows increasingly difficult as they move up in the grades. Play, once a serious activity for investigating the world and uncovering some of its mysteries, becomes relegated to the *entr'acte*, a sort of coffee break for child laborers. It is not that children need unrelieved play. Far from it. With their world hidden from view, and with few skillful adults to observe and mimic, the monotony of everlasting play soon puts children in sympathy with the child in a famous cartoon who said, "We're tired of doing as we please, can't we do something else?"

Play as an aspect of the study of science gives children the opportunity to assess the nature of materials and to consider the range of questions and problems the materials suggest, the possibilities for change and novelty. Although carefully structured in that precise materials and equipment are made available at any one time, play activity is undirected, leaving the child's imagination free to roam, unencumbered by a fixed focus, free to choose, to make independent decisions and observe their consequences.

To acknowledge play as a necessary but insufficient condition for learning science is consistent with what scientists tell us about the nature of their work. It is perfectly respectable. Einstein maintains in '*On the Method of Theoretical Physics*' (1934) that "all knowledge of reality starts from experience and ends in it," and in *Atomic Physics and Human Knowledge* (1958), Niels Bohr states, "all knowledge presents itself within a conceptual framework adapted to account for previous experience." Play provides the raw experience from which concepts and knowledge will develop. P.B. Medawar, an English biologist who won the Nobel Prize for medicine in 1960, notes in *The Uniqueness of the Individual* (1961): "So far as I can tell from my own experience and from discussion with my colleagues, hypotheses are thought up and not thought out. One simply 'has an idea' and has it whole and suddenly, without a period of gestation in the conscious mind." It is among the functions of play to generate hunches and ideas. Play will fulfill these functions if it engages children in activities that are interesting, significant, and open-ended.

RESPECT FOR INDIVIDUAL INTEREST

The test of the effectiveness of the play episode in science learning is whether the criteria for the selection of materials and equipment—interest, significance, and reality—are actively fulfilled for *individual* children. Consider children at play inflating balloons—watch what they do and listen to what they say to each other. Some tie the inflated balloons off and hit them from below, driving the balloons almost to the ceiling. They watch as the balloons slowly descend.

"If it was the kind you get in the park, it would just stay up there and never come down."

"No, it wouldn't. I brought a park balloon home, and it got smaller and smaller, and quit."

"Must've had a leak."

Someone else is squeezing a balloon, enjoying its springy quality. Another child decides not to tie hers off, but to let it go careening through the air. The idea catches on, and a friendly competition follows to see who can get a balloon to travel the entire distance of the room. However, no balloon maintains a straight path; some even double back and go the "wrong" way. Two children try to see how large they can make their balloons. After the balloons have exceeded their usual size, the children look at each other in cheerful but nervous expectation. One child rubs a balloon against the sleeve of her blouse and rests it against the wall while her neighbor announces that they always do this at her birthday parties. Everyone is startled by the noise from a "popped" balloon. When all have been suitably impressed with the volume of the sound, a lad who probably regretted that he had not been the first to break his balloon assures everyone that he can make a louder bang with a paper bag.

Play

These are all common experiences with balloons. One may reasonably expect that children will engage in them and in others of their own invention if they are provided with balloons and time in which to try, and if no one intrudes. Left to their own devices, one or two children may pursue a puzzling feature of their experience. Perhaps someone has seen an illustration of a balloon traveling along a string and may try to repeat it. Some children may try to attract each other's balloons by rubbing them on various materials. Most children, however, will eventually tire of these activities or, more likely, break the balloons and look for something else to do.

The play must not be allowed to die of exhaustion. While only a few children may have been launched into activities of their choice, others may need a more direct challenge to come to grips with the problems that remain unrecognized. (What they do *not* need is an analysis of the nature of air.) A jar is introduced with a balloon partially inflated inside the jar, and the open end of the balloon stretched over the mouth of the jar. There is another jar with a waterfilled balloon wedged in the mouth of the jar, most of the balloon outside, and part of it inside the jar. Another inflated balloon rests inside a funnel which extends out of a length of tubing. The other end of the tubing is submerged in a pail of water. When the funnel is turned upside down, the balloon does not fall out. It requires some effort to pull it free. These are some of the puzzling things that one can do with balloons. They invite the children to try to do them, like tricks.

The play is now somewhat directed. The adult has intruded, but he has done so by turning the direction of play in order to raise the level of interest, heighten the significance, and yet leave open the reality that may be discovered. There are as many ways to do this as the adult's imagination and experience suggest.

He may show how big he can make a balloon with one expulsion of breath. It is a measure of his lung capacity. What is theirs? How is such capacity to be measured? How do the capacities of the children compare? They can measure the circumference of the inflated balloons with a tape measure or string, determine the volume however they can, and record the results in line and bar graphs: lung volume against individual children, lung volume against circumference, lung volume at rest compared with after exercise, lung volume against chest measurement, or whatever other relationships may be of interest.

The adult may play on his primitive recorder made from a drinking straw. I have never met a child who did not want to make his own. "Do you suppose it makes any difference where you place the holes?" "Will it sound different if it is shorter?" "Is there a way to make one so that the scale will be the familiar one instead of the strange one we have produced?"

If some aspiring rocketeer has brought a carbon dioxide cartridge to show how fast his model travels across the room along a wire; that is fine.

CHILDREN CHOOSE THEIR ACTIVITIES

It is not intended that all the children play with balloons at one time. It is physically inconvenient, and it is likely to be confusing. In any case, children should be able to exercise choice even at play, or especially at play. There are a number of other activities available to them. Among these may be a pendulum.

A pendulum at rest invites a push, just as the sign "wet paint" invites a touch. The pendulum may be suspended from a simple wooden frame, and a can partly filled with sand can serve as a bob. There are golf balls, softballs, marbles, and rocks for alternate bobs, and a variety of cords and thread. Available, too, are the feet of old nylon stockings to contain the bob, a simple arrangement for children not yet able to manage attaching a string to a spherical surface without assistance. There are also laths, rulers, and yardsticks. One yardstick has a nail driven through one end, and it is balanced with the nail lying across two yardsticks which rest on the backs of two chairs. It is a rod pendulum. To know what needs to be

Play

taught, and to find the teaching moment, the children are observed. Who is measuring the pendulum? How do they determine the length when the bob is regular? Irregular? Who is seeking a relation between the period and the amplitude? Between the period and the weight of the bob? Is anyone comparing the string pendulum with the rod pendulum? Is anyone using the books that were carefully selected to accompany this activity?

The play with pendulums should be made increasingly complicated, but not too quickly. After children have become absorbed by the simple pendulum, coupled and compound pendula may be introduced, raising the interest a notch or two. This is not laid down as a general thesis about going from the simple to the complex. Often children do better when confronted by confusing but interesting problems first. In this case, however, once the children are offered a compound pendulum, especially one capable of tracing the lovely curves called Lissajou figures, which are produced by the interaction of two harmonic motions at right angles, their satisfaction in generating these beautiful curves in sand or paint may dwarf the pleasures of the simple pendulum, and that would be a pity.

The strong recommendation that play serve to initiate scientific activity is not to be interpreted as a precise judgment for every child at all times. It is sometimes possible to discern distinctions in style and temperament among children in their attack upon science problems, especially once they have learned to read and to perform useful mathematics. Although eventually I should hope to interest as many children as possible in such things as gravity estimations and the powerful tool put at our disposal by sentences which relate time and the length of the pendulum, occasionally a child may not want to wait. One needs to keep a sharp eye out for the youngster who gets his pencil and pad out as soon as he observes the results of manipulating the equipment. Such children are a small but precious minority. They usually want books and to be left to their work.

The distinctions in style among children are analogous to those we find among scientists. J.B.S. Haldane, a renowned English biologist of this century, was the first man to map a human chromosome and measure the mutation rate of a human gene; he was genuinely interested in theory. Scientists so disposed seek the implications of theory and proceed to devise experiments to test their ideas. Haldane performed many remarkable experiments, even on his own body, to test the validity of theoretical propositions. C.V. Boys, whose lectures on soap bubbles were mentioned earlier, was temperamentally very different from Haldane. Boys generated ideas in the very process of working in the laboratory. When he was offered the services of a mechanic to assist him in the construction of apparatus, Boys refused to accept the help. It was in the very construction of apparatus that he often formulated his ideas. Children may display similar differences in the way that they prefer to work.

THE OUTDOOR LABORATORY

Learning about the remarkable variety of living things requires a playful, undirected opportunity to become acquainted. The scaled and clawed collared lizard resting on the hand distinguishes himself and his kind from the toads, frogs, and salamanders. The eggs that appear in strings, masses, and spotted clusters give promise of similar yet distinguishable creatures to come. The bluffing hog-nosed snake and the aggressive milk snake soon adapt to gentle treatment. The box turtle hides completely in his shell, but the snapping turtle does not need to do so. One tree frog seems to have disappeared on the bark while another is resting comfortably on the glass of the terrarium. No models, however ingenious, are a substitute for live animals and plants. Models and pictures are useful in addition to, not instead of, the many interesting organisms that can reside among children.

The field is a source of many of the living forms in whose company children may begin to marvel at the variety, the similarities

Play

and differences, the adaptive mechanisms, and the interdependencies among plants and animals. Interest, significance, and reality abound at every hand; for the children to discover them requires only time, patience, and sympathetic but unobtrusive guidance. The decaying log and its immediate neighborhood is a little world related to other little worlds in its vicinity. Among the fungi penetrating the remaining woody tissues are springtails which are perhaps a quarter of an inch long. Later, much later, the children may put their knowledge of large numbers to work to see whether it is really possible that there may be 100 million springtails per square meter in some soils. Later also, they will fashion a Berlese funnel from a conical paper drinking cup, their soil sample inside the cup, a piece of nylon screening across the open bottom of the cup, and a 100-watt bulb hanging over it. A jar of alcohol placed underneath will catch the insects that burrow through the soil and the screening as they try to escape from the heat and light of the bulb. But for now the children are observing and collecting.

There is a millipede, curled up like a watchspring, exposing the part of him that feels hard when he is picked up. There is a sowbug, cousin to the lobster, crab, and shrimp. The centipede runs off on his many legs. Are there really a hundred of them? The children will have to catch one to count them. He runs when his head is uncovered. That is understandable. But he also runs when it seems that his head is still hidden. That is odd. Is it true that the centipede will stop running if a vessel is provided that will allow his body to make maximum contact with a surface? Perhaps some tubing would be best. The snails are very ancient citizens, having left their fossil tracks some 500 million years ago. It may be more interesting to collect several in order to find out whether snail mates really shoot darts at each other in their gastropodic romance. The children need not worry about including male and female snails because each snail is both.

Those seemingly helpless slugs have been known to drive birds off with a squirt gun they possess. How slimy to the touch they are. Back in the school library is a picture W. La Varre took of a slug

Teaching Science to Children

gliding over the sharp edge of a blade without harm.

Judging from the cloud of spores traveling in all directions, I should think that someone must have broken a puffball open. How many tiny spores there must be in these kissing cousins to the mushrooms. The children add a puffball to their collection. Perhaps some brave soul will try to estimate the number of spores contained in the fruiting body. Some bracket fungus is added to the collection so that later the children may examine the lower surface with a hand lens, see the tubes where the spores are developed, and explore the role of this fungus in the life cycle of the woods. Now the search is for inky-cap mushrooms—the children will try to write with their ink—for edible mushrooms, which will not be eaten until an expert determines that it is safe, for varieties of cup fungi, and especially for morel which is good to eat. Samples of lichen, that marvelous partnership between alga and fungus. the pioneer of the plant kingdom, will be collected. Finally, after ferns, liverworts and mosses have been added, it is time to move on.

Before the children leave the school in the outdoors, they still have much to do. Last night, before bedtime, they listened to some poems. Among these was one about spiders by Robert P. Tristram Coffin from his *Collected Poems* (1929):

> He lays his staircase as he goes
> Under his eight thoughtful toes
> And grows with the concentric flower
> Of his shadowless, thin bower.

The children want to find at least one orb-weaving spider. We have built a special frame and want to watch as it weaves its web. They also want to find at least one variety of ant before moving on to a new community—the pond. The social ants, whether food gatherers, farmers, stock breeders (the hunters will be left alone), will challenge the children's understanding of mankind's unique cultural invention. In what ways are the ant communities similar to human communities? In what ways are they different? Perhaps

Play

some child will be fortunate and come upon harvesting ants among which "we find an analogue of beating swords into ploughshares; the 'soldiers' with their vast heads and powerful mandibles, crush seeds for the benefit of the whole colony." This lovely description by S.A. Barnett in *Instinct and Intelligence* (1967) stimulates the search.

The opportunities to observe and collect at the pond are too numerous to list in any detail. The children find dragon, damsel, may, and caddis flies at various stages of development and see water striders, water boatmen, backswimmers, whirligigs, diving and water scavenger beetles, and the ubiquitous mosquito. Growing in the water are waterweed, bladderwort, water silk, and other plants. Some of the animals and plants are collected for the aquaria at school. They also take along some of the pond water to keep the animals and plants alive and plan to examine it under the microscope. After all, the food chain in the pond may include plants and animals too small to see with the unaided eye.

The meadow presents new opportunities. What varieties of life can be found under two square feet? How plentiful are earthworms? While collecting a number of worms for their worm garden, the children try to estimate the density of worms in a given volume of soil. Their estimate is guided not only by Charles Darwin's count in *The Formation of Vegetable Mould* (1881), but by the worms' work as measured by the quantity of castings they have produced. Other children may prefer to consider the secret of the dandelion's success. There were none to be seen in the shady woods, but here in the meadow they are well established. It is interesting that people clear the land when they settle in a new place and then complain when dandelions find the disturbed land precisely what they need. If changes were to be made in this outdoor laboratory, what likely consequences could be anticipated? For example, what might happen if the pond were filled in? Or suppose that the woods were cleared to make room for more homes?

Until now everyone has been looking down most of the time.

Now it is time to look up before returning to the indoor laboratory. Each of us wrapped against the night chill, we lie on our backs and gaze into a sky most of us have not seen; we wonder, dream a little, and ask a question or two, feeling subdued and awed by the darkness, the silence, and the immense sky. *Bolts of Melody, New Poems of Emily Dickinson* (1945) includes "Once a Child," a poem that expresses feelings some of us share:

> It troubled me as once I was,
> For I was once a child,
> Deciding how an atom fell
> And yet the heavens held...

PLAY IS A BEGINNING

When young city children go to the countryside as part of their science experience, they need time—time to run, run, run, without meeting obstacles in walls, traffic, people. They need time to throw stones, something one cannot do in the city without inviting reprimand. They need time to splash at the edge of the pond and to build intricate canals in the adjacent sand. When they have greeted the new environment, their teachers can guide them gently to a small sample of the myriad life forms that have made each little niche a home to sustain life. It is not enough to let children loose: they see too much and, therefore, perceive very little. They may not think to lift the bark, or scrape the "paint" from the rock, or note the dimple in the water under the water strider's foot. Caesar had to be turned only to hear; children have to be turned to see, to smell, to touch and not to touch, and to taste and not to taste.

When the opportunity to explore freely is offered to all the children, some will do a little and others more. Some will find themselves in dead ends, and they will learn to turn themselves

Play

around and try another route. Others will be mistaken about many things. No matter. Edwin H. Colbert's advice, given in *Men and Dinosaurs* (1968), urges forbearance in making judgments about the fumbling efforts of the pioneering paleontologists and suggests an attitude toward people who are beginning new experiences: "In paleontology, as in other branches of science, and for that matter in other branches of human endeavor, something is better than nothing, and it is well to do something with that something. If men were to wait for complete knowledge before attempting anything . . . little would be accomplished." The play episode is a beginning. It will be followed by different episodes.

So far the children have been given an opportunity to play in environments chosen by, and in some cases designed by, adults. There has been no effort to encourage children to pursue the adult's linear thinking. Instead, provision was made to encourage the storing of varied memories and experiences so that these might allow thoughts to branch out naturally, leading to new associations and ideas, making room for imaginative vision. Rigid ideas, suffocating erudition, and bad teaching have been absent, and so their deadly effects have been avoided. Support has been directed to each child in his efforts rather than to a project of one kind or another; it is similar to the support given to a scientist doing basic research and unlike the support given a project in applied research.

The play has been free, its nature largely determined by the child's inner response to the materials offered him. Children know that free play, although necessary and delightful, is limited. They seek new means in their explorations of their capacities to deal with the world. As children's play becomes more social, it is increasingly governed by rules, some of which have been passed down, while others are invented. Autonomous play has been converted to social games.

3. Games

There was a child went forth every day,
And the first object he look'd upon, that object he became,
And that object became part of him for the day or a certain part of the day,
Or for many years or stretching cycles of years.

Affection that will not be gainsay'd, the sense of what is real, the thought if after all it should prove unreal,
The doubts of day-time and the doubts of night-time, the curious whether and how,
Whether that which appears so is so, or is it all flashes and specks?

These became part of that child who went forth every day, and who now goes, and will always go forth every day.

Walt Whitman, "There Was a Child Went Forth"

Most people are bored by their work. Work has been growing less interesting since the early part of the eighteenth century when new sources of power enabled English industrialists to divide labor into a series of manual operations in the course of which the former craftsman became a specialized implement, a production mechanism. The new organization of production proved remarkably efficient. It relieved the workman of the need to consult his mind, eliminating reflection and fancy, and in doing so it speeded up the work and excluded error. By the middle of the eighteenth century some English manufacturers who had operations they did not want to divulge sought to hire seriously mentally retarded people

Games

to do their work. This was entirely logical. A person who spends his working life performing a few simple operations that do not require him to exert his understanding is stupified by the experience. One does just as well to hire people who are already stupified.

For the majority of people in industrialized nations, work has been a dreary experience ever since industrialization occurred. Many people try to save their intelligence, their imagination, and their sanity through various activities after work. That is when life really begins for them. Meanwhile, they try with whatever means are available to them to reduce the hours of work required to earn a livelihood. Some few still share the dream, at least in America, of opening a corner garage or a small store. But the best thing is to avoid work altogether—an achievement beyond compare and the surest mark of success.

Though for most people work is a socially imposed burden which they would gladly shed if they could, a minority find their work fulfilling. It gives their lives meaning and purpose. From 1898 until 1902 the French chemists Marie and Pierre Curie purified and purified again tons of waste ore rich in uranium to obtain, at the end of this grueling procedure, one tenth of a gram of radium. Einstein reports, in the essay cited earlier, that between 1912 and 1914 were "two years of excessively hard work." He was then developing the General Theory of Relativity. The work of these scientists was burdensome, but certainly not unwelcome. They throve on their work, it was the central meaning of their lives.

This attitude toward work is certainly not restricted to scientists. Think of the frenzied concentration of Van Gogh at Arles. Imagine *choosing* to arise at four o'clock in the morning to begin the day's work. We have it on Stephen Spender's authority, as expressed in *The Making of a Poem* (1946), that for poets "everything is work except inspiration." These scientific and artistic workers and their colleagues labor in a tradition far older than the factory system. They are continuous with Michaelangelo, who spent four years

Teaching Science to Children

painting the ceiling of the Sistine Chapel; with Cervantes, poor, desperate, and three times imprisoned, who nevertheless spent fifteen years writing *Don Quixote;* and with Handel, who in twenty-four days of furious work wrote the *Messiah.*

In this tradition, as in the ancient craft tradition, the worker's imagination seizes some portion of reality—rock, space-time, color, language, sound—and imagines how it may be changed. He enters into a contest with it and works until he is satisfied that he has most nearly brought about what he had envisioned. The factory system has made incursions upon this tradition, but it has not yet been able to destroy it.

The scientific contest with the environment, like the other traditions, is played according to rules and reveals in its results the skill and good fortune of the contestant. This is not at all an adequate description of what transpires in scientific work. But such work does contain these elements, and the activity, when limited to these characteristics, serves as an analogue to games, something children know very well. Games with rules capture some of the essential spirit of children engaged in scientific activity. It is the kind of game theorists describe as non-zero-sum, where the gain of one player is not the loss of another. The contestants are the natural environment, which displays countless apparently disparate particulars and challenges the other player, the investigator, to isolate some fragment to find pattern and unity. There are rules to be observed, but their observation no more guarantees success, discovery, than obeying chess rules guarantees a winning game. The game must be sufficiently interesting, or children will engage in it only reluctantly and will not employ their imagination, skill, and persistence. Worse, they may not be engaged at all.

The game metaphor must not be strained. It would be absurd to think of the teacher or another adult as a referee, the findings as scores to determine standings in the league, and Nobel prizewinners as international champions, although science fairs, it is sad to say, often come very near to a Little League competition. My intention in suggesting games as a fitting metaphor for children's

Games

explorations in science is to militate against a popular view that to learn science one must follow prescribed routes. It is no more true that one discovers in science by following the route from hypothesis through observation, data collection, experimentation, to conclusion, than it is that one writes felicitous letters by beginning with a salutation and proceeding through exordium, narration, and petition to a conclusion. More suggestive of the liveliness in science work is a description by Percy W. Bridgman, the American physicist who received the Nobel Prize in 1946, which is quoted by Abraham Kaplan in *The Conduct of Inquiry* (1964). Bridgman remarked that "the scientist has no other method than doing his damnedest."

Pitting oneself against the puzzles of natural phenomena in an effort to extract their secrets is an activity with gamelike qualities. Some government officials have, in the past, resented this joyous characteristic of basic research, and have foolishly used it as a reason for denying funds. Such people fail to understand that it is this very quality, the self-motivated pursuit of interesting questions and problems, the challenge to resolve enigmas, that has enabled scientists to be so remarkably productive. It is odd that people who themselves are often captivated by tangrams, Chinese blocks, "Instant Insanity" blocks, and similar puzzles, do not comprehend that others may be similarly intrigued by nature's own puzzles. A good way to gain an understanding that the difference between the common fascination with man-made puzzles and the scientist's fascination with natural ones is a matter of preparation and focus is by reading from the history of science.

A SCIENTIFIC PUZZLE

In the preface to the Fifth Book of *Harmonice mundi* (Harmony of the World), which Johannes Kepler, the great German astronomer published in 1619, he wrote: "Having perceived the first glimmer of dawn eighteen months ago, the light of day three months ago, but only a few days ago the plain sun of a most wonderful vision—nothing shall now hold me back. Yes, I give myself up to holy raving. I mockingly defy all mortals with this open confession: I have robbed the golden vessels of the Egyptians to make out of them a tabernacle for my God, far from the frontiers of Egypt. If you forgive me, I shall rejoice. If you are angry, I shall bear it. Behold, I have cast the dice, and I am writing a book either for my contemporaries, or for posterity. It is all the same to me. It may wait a hundred years for a reader, since God also waited six thousand years for a witness." Now, of course, Kepler cared very much whether his contemporaries, Galileo, for example, valued his work. But I believe the sense of challenge imparted in these words is true. Few scientists are as effusive in our less passionate times, but the drive that spurs their work is no less intense. The scientist's work is not cold and mechanical. An account of Pasteur's work; of the many scientists who unraveled much of the atom's structure; of the work on proteins by Linus Pauling, a contemporary American chemist who twice won the Nobel Prize; and of any number of other scientists would be equally persuasive that when "the game's afoot," they too follow their spirit, but not, let us hope, for England and St. George.

The story of the German Friedrich August Kekulé is not unusual. The historical and personal circumstances would be different, but the quality of the contest between natural phenomena and man's imagination, ingenuity, and persistence would not be unlike that in other stories that might be told.

Ethyl alcohol, the kind people have been making since ancient times, the alcohol we drink on social occasions, is composed of molecules, each of which contains two carbon atoms, six hydrogen atoms, and one oxygen atom. The molecules of dimethyl ether are composed of precisely the same atoms in precisely the same numbers. That is, one formula can represent the atomic composition of both these substances. Yet, dimethyl ether is a gas under normal conditions, and drinking alcohol is a liquid under normal conditions. Substances which have the same molecular formulas, but whose properties are different, are called *isomers,* from the Greek word meaning "equal proportions." How can one account for the existence of isomers? If the problem were restricted to its appearance in ethyl alcohol and dimethyl ether, it would still be interesting, but of limited scope. However, the number of possible isomers increases rapidly as organic molecules, those containing carbon atoms, become more complex. There are, for example, thirty-five different forms of the molecule composed of nine carbon atoms and twenty hydrogen atoms. Most organic compounds cannot be identified simply by the number of each kind of atom. It is not difficult to understand that this should be so. The number of each kind of letter on this page is not sufficient information to know what is said on the page. In order to know that, we need to know how the letters are arranged.

This was the puzzle to which Kekulé gave his attention more than a hundred years ago. He decided that carbon atoms, which appear in all organic substances, could be attached to other carbon atoms to form chains. A new way of representing compounds became commonplace in chemistry. The new representation not only gave the atomic composition, but also the arrangement of the atoms. It was a representation of the molecule's structure. Now it is less surprising that ethyl alcohol which can be represented as

$$\begin{array}{ccc} H & H & \\ | & | & \\ H - C - C - O - H \\ | & | & \\ H & H & \end{array}$$

exhibits properties which are different from those of dimethyl ether, which can be representd as

$$\begin{array}{ccc} H & H \\ | & | \\ H - C - O - C - H \\ | & | \\ H & H \end{array}$$

Kekulé was not finished yet. Open-chain formulas of the kind illustrated, and other, hypothetical formulas, did not account for the properties of benzene. Nor do such formulas account for the isomerism of benzene derivatives. Kekulé introduced the idea of *rings* of carbon atoms giving benzene the structure of a regular hexagon. A hundred years later, X-ray data and electron diffraction studies confirmed the correctness of the hexagonal feature of benzene molecules. Kekulé had won, but it was more than a personal victory. The synthetic dyes which were then beginning to be produced could now be produced more rapidly.

The Dutch physical chemist Jacobus Henricus Van't Hoff and the French chemist Joseph Achille Le Bel added a third dimension to Kekulé's two dimensional structures. Early in this century the American chemist Gilbert Newton Lewis elaborated an electronic structure; later, Pauling developed knowledge about molecular structure still further through the use of quantum mechanics. The game is played like a relay; each participant passes the baton to the next one.

MODELS

In the course of their absorbing, self-directing work, scientists frequently employ physical models. Models make available on a manageable scale an analogue of some portion of our universe that is either too large, too small, or too intractable. Whether it is the Copernican model of the solar system, the Bohr model of the atom, Pauling's model of protein, or the Watson-Crick model of DNA, the model is designed to embody structurally an analogy of the physical world. If they are good models, they contain patterns which correspond with those in the real world that inspired them.

In building the model, all the knowledge that has been acquired serves as the starting point, the rules by means of which it is visualized. Copernicus began with Ptolemy's epicycles and reduced them, simplifying the model. Bohr inherited the atomic model of the British physicist Ernest Rutherford. Pauling had Lewis before him, and Watson had Pauling. Even Kekulé did not

Games

make up all the rules himself. Like other innovators, he changed some of the rules, but he had a long tradition behind him which extended back all the way to alchemy. Nearer at hand was the work on valence of the Italian chemist Stanislao Cannizzaro and the English chemists Alexander W. Williamson and Sir Edward Frankland. Like other creative work, science invites finding the rules in some way inadequate. It demands as a condition of joining the scientific fraternity that a member improve upon them. To do that, all work must be available for inspection. There is no secret science.

SCHOOL AS A LEARNING LABORATORY

One of the central mistakes of the school has been to model children's learning experience, including their science learning, much more nearly after the factory than after the real activity of scientists and other creative workers. Children report to an assigned place in an assigned room in an assigned building. They are given work to do, and the product of their labors is then inspected by a sort of product control mechanism. After the work has been inspected, the children are adjudged fit or unfit for promotion. At no time must they incur the displeasure of the foreman (usually a forelady) or risk suspension from the job. At the end of stated periods they are given chits to take home, more or less adequate to sustain their self-respect. Of course, this is a gross overstatement, and it is certainly not what many dedicated teachers would want.

Many children, particularly in the intermediate and upper grades, do not find the factory description of the school an amusing caricature. The quest for efficiency is not funny. It is deadly serious Education has become big business. What is more natural, when education is to be "businesslike," than to invite system analysts to help get out a good, standardized product, "a nice package," at minimum cost in time and money. Such a system may

Teaching Science to Children

produce neat packages, but it will not educate children. Like the old factory system, it will only stupefy.

The model for organizing an environment fit for children is not the factory but rather the laboratory and the art studio. Here children can pit their minds, hands, and spirit against natural phenomena, which only reluctantly divulge their secrets. An administrator of the laboratory conceives the strategy. He plans the program with the children, provides them with materials and books, and gives guidance and encouragement. He leaves the tactics to the children.

In the games played in the laboratory there is room for the most diverse talents, skills, personalities, and temperaments. Some prefer to work alone, the old way of science. So be it. Others prefer to work in groups, the new way. Some of these groups will occasionally be "nodding committees" in which the leaders suggest what is to be done, and the others nod in assent. That too is a way of learning. Players on a team are rarely of equal skill. Weaker players gain strength in association with more able ones. If the leader mistakes his role and behaves as a foreman toward laborers, the members of his group will soon correct his error. If he proves too disagreeable, he may find himself working alone until he learns to behave more acceptably. It is a lesson worth learning, and it is best taught by one's colleagues. However, a pleasant disposition is not a condition for participation in science. If it were, the seventeenth-century English physicist Robert Hooke and the eighteenth-century English chemist and physicist Henry Cavendish, among others, might have been excluded. That solitary figure who is chipping flint to make weapons and tools in the corner may indeed be disagreeable. We hope that he will become less so, but meanwhile he can learn science by playing his game alone. If the poor fellow is condemned to suffer his indisposition, it is better that he suffer it as a science worker than as a teacher. The September 1969 issue of *Scientific American* carries an advertisement inviting resumes from scientists, "Especially if you're the disagreeable type." I have yet to see a similar invitation to teachers.

Games

Not only people, but things, can be unpleasant. A pumpkin, having seen its days of glory, is now reduced to a soggy, misshapen, discolored mound. The appearance of a strange, furlike growth, and its odor, disgusts many children. But not everyone is inclined, as Francesco Redi was, to use decayed meat and fish from a variety of animals to do embryological research. Nor is it to everyone's taste to use pus from surgical wounds, as the Swiss physiologist Friedrich Miescher did during the last century, to study genetics. It is sufficient that some are more fascinated than repelled by such opportunities. The pumpkin is, by common consent, an ugly mess, but in a laboratory such a mess is also an interesting object to investigate. It is not the function of the director, whatever his own tastes may be, to remove it in order to maintain a sterile environment. On the contrary, he struggles against sterility.

A small group of children is interested in the strange growths that have infected the pumpkin. They examine it at a distance, then closer, and finally with the help of a hand lens. They speculate about what the "furry" growths may be. Someone is reminded of similar growths on bread and oranges. They organize a campaign to track the mystery down. The children are transformed because they have found something that stirs them to action. Textbook assurances about the characteristic attention spans of young children go by the board. Homework is self-assigned, and parents complain about finding jars with bits of food in all parts of the house—the refrigerator, the closets, the bathroom, and on radiators. All is fair, not only in love and war, but also in scientific games. Records are kept of the kind of food, its condition at the start, the place where it was kept, how long it remained there, and what changes were observed. There is a run on books with references to molds. Now the game is almost over. It remains but to share the experience with everyone who may want to know and to post the results.

The "colloquium" will be less imposing than that at the Cavendish laboratory under Sir Joseph John Thomson, but it will try to capture some of that atmosphere of genuine scientific interest combined with social affability. That Thomson was a great scientist and was awarded the Nobel Prize in 1906 for his work on the electron does not fully account for the remarkable fact that seven of his assistants subsequently won Nobel Prizes. I suggest that those sharing seminars may have made an unusual contribution to the cultivation of talent. In our humble effort to emulate the atmosphere of interest and affability at the Cavendish laboratory, we may go so far as to copy Thomson's procedure and use the time immediately preceding the presentation of the completed work and the following discussion for a brief social respite, a kind of seventh inning stretch. Though there will be no ladies serving tea, children can serve cookies and a drink. It is the rare child who excludes himself from the snack, the casual conversation, the laughter and banter of the social occasion that sets the stage for the sharing experience.

Everyone who may be interested is invited to come. Those who cannot be present to hear the personal accounts of the investigators may read their short written accounts, or they may seek out the researchers at some future time. This is the way of science. Scientists do not keep their findings to themselves—they announce them. That is how scientific knowledge grows.

For example, when a man's body prepares for an emergency, his heart pumps faster, he breathes faster, his muscles get ready to consume energy faster, and he is less sensitive to pain. Today physiologists of the human body know a good deal about the hormones, the chemical messengers, that are responsible for these and many other changes that take place in the human body. Many scientists contributed to this knowledge, including the biologists who studied the hormones in insects. If Pierre Lyonet (Dutch), Sir Vincent Wigglesworth (English), Stefan Kopéc (Polish), Soichi Fukuda (Japanese), Peter Karlson (German), Carroll Williams (American), Karel Sláma (Czech), and Kailash Saxena (Indian) had each been content with his private knowledge and had not announced his work publicly, much less would be known about in-

Games

sect hormones today and knowledge about the chemical messengers in human beings would be severely retarded.

Children need the experience of this kind of collaboration. While the children's presentations affirm that investigation is not science until it is shared, they also serve to reveal the extent to which children have gained insight into the art of investigation. Also, public presentations leave open the possibility that a listener may find that the work described sheds light on his own work. Finally, confronting questions posed by one's peers conveys an intimate conviction that to tell the truth in matters of science is the only sensible thing to do. Dishonesty is rare in science, not because of private virtue, but because of public accounting.

While the group working on molds was preparing to give an account of itself, another group has been building a model of the community. Children are very fastidious about models, which is why they prefer dolls that "talk" and "wet." It is also the reason that they find flat maps so puzzling. The attitude they display in building their city is very much like the one seen in children who are building castles on the beach. The principal difference is that cardboard buildings on a wooden floor are more durable than sand castles on the ocean-washed beach and therefore permit greater detail. The children insist that the buildings must have electricity, water, and sewage disposal; that the streets must have street lights, traffic lights, and waste baskets. There are problems. Electric bells must be eliminated because they are too large. When a bulb burns out in one of the buildings, the city suffers a blackout—back to the drawing board. Books prove their usefulness. There is apparently more than one way to hook up a circuit. Are the buildings to control their individual lights? How about each floor?

Water presents new problems. How big should the reservoir be? Where should it be placed so that all the floors will have water? If the reservoir is to be made of plaster, it will be very heavy, especially with the water in it. Is there a way of deciding in advance where it should be placed? Back to books, trials with a funnel and a length of tubing, until a decision is reached. By this

53

Teaching Science to Children

time there is a considerable pool of water on the floor. What is one to do? Creative work is bound to be messy and destructive. So it goes until the children are pleased with what they have built. And because they are pleased and confident about what they have done, their presentation is not an ordeal by verbal fire. Neither is it an exhibition of superiority. The children's presentation is a straightforward account of their purpose, procedures, difficulties, errors, readings, assistance received, and their qualified assessment of the results.

Although it appears during the presentations that each member of the group has contributed significantly to the results, the administrator knows that this is not so. Not everyone in a group exerts his efforts to the fullest extent. Some do not do very much or very well. We need not, in every case, paint gloomy futures for them. The future of Jöns Jacob Berzelius, the Swedish chemist who dominated the field in the early nineteenth century, had once been very much in doubt. There are times when children indulge in fantasy. Remember Darwin and Tsiolkovsky? There are times when children seek the physically stimulating activities. Perhaps, like Ivan Petrovich Pavlov, the Russian physiologist who died in 1936, they need, at these times, to feel glad in the muscles.

Despite the variety of preferences that the children may exercise, there may still be some children who appear to do nothing at all. None of the choices open to them is appealing. They do not want to assemble the bones found on a field trip in the country or to build a model beaver lodge, or a compound pendulum to trace lovely designs, or toys for the bazaar. Nor do they wish to measure the relative strength of various shoots as they push their way through the soil, or measure the diameter of the sun, or build a sundial, or make dyes from plants, or any number of other things they might do. They prefer to remain inactive. The laboratory director observes them. If he insisted that every researcher must be constantly busy, we should be right back in the factory again, or in the Army awaiting inspection. If observation and familiarity with the pattern of these children's behavior suggest that their

Games

inactivity is a pause, a refreshing interval, they are left to their daydreams, their fantasy, their muscular play, or small talk. It is not unusual for children to remain out of a game for a time, and then, when the spirit moves them, or upon invitation, to join again. The situation is similar here. However, something needs to be done for the children who are not fired by the enthusiasm that grips the various groups or by ideas that they might pursue on their own. These children are not successful self-starters; they need a push.

Task cards on which problems are already formulated meet this need. The problems may be in the form of puzzles, constructions, or directions for conducting an experiment. Included among the activities which the cards encourage are some that are related to the ongoing activities in the laboratory. There are puzzles with circuits in series and in parallel, puzzles based on water pressure, and other puzzles that may provide an interesting contribution to the work of a group. Perhaps the child is reluctant to join a group empty-handed or empty-headed. The cards may provide him with an offering.

There are other tasks involving the construction of materials especially useful to younger children. Kindergarten children can always use more mystery boxes. These are sealed containers of various sizes and shapes which hide something. Whatever is hidden in the boxes reveals its presence only by the imaginative use of the senses. For example, there is a box with a xylophone hidden inside. On one side of the box are several slots large enough for a washer to pass through. The child who thinks of slipping metal washers or coins through the slots may guess from the different tones what the box contains. Another box has a hole on one side. Careful observation reveals that there is a string or wire inside. A pencil which probes through the hole against the wire or string, elicits a musical tone. There are other containers for smelling and for touching, and still others for tasting. There are never enough of these mystery boxes. Some children who feel inadequate with their peers gain immense satisfaction from helping younger children. Constructing these boxes gives them an opportunity to do so.

Teaching Science to Children

Finally, there are cards with random activities. "Can you construct a minute pendulum? How many times each minute will the pendulum swing if you made it half as long? A quarter as long?" "Here is a way you can move a coin from under a glass without touching coin or glass." "You can make a paper 'roof' turn by just bringing your hands near." "Here is an experiment you can do to see what happens to weight during free fall." "You can build a thief-proof drawer." There are many such cards. They are attractively made and plastic-coated.

The children who have not been included in scientific activity either by their own choice of problems, by associating with someone else, or by finding a suggestion in a book or a task card, present a special problem which cannot be assayed without knowing the individual children. Even then, it is often very difficult to decide why they exclude themselves. If they participate in other activities, it is not serious, and usually something can be done. This kind of selective exclusion is rather rare in mhisy experience. More common is the child who does not participate in a wide range of activities. Their interests may be exclusively intellectual, physical, or artistic. Withdrawal is a symptom of many kinds of underlying problems, and it requires the help of experts. The child who remains outside the laboratory games despite every effort of the director usually requires the attention of such experts. Their help should be sought.

THE TEACHER AS LABORATORY DIRECTOR

While the laboratory is at work, the director makes the rounds about the laboratory. He observes and listens very carefully, especially to what children say to each other. In these exchanges he may find clues to what ought to come next, what changes ought to be made in the environment. He is occasionally asked a question as he moves among the children. Some questions are simple requests for a straight bit of information: "Who first thought of doing

Games

all these things with a pendulum?" "Can we get balloons that we can blow up to 6 feet across?" "Where can we get mealworms for our frog?" If he knows the answers, the director ought to give them. The child who wants to know such things may think of little else until he finds out, and they are not worth the effort or time he will expend in finding out. If the director does not know, he ought to say so and suggest possible places where he and the child can look for the answers.

But there is another kind of question. "Where do you measure when you measure the length of a pendulum?" "How come the balloon gets big all around when I blow only one way?" "Why doesn't the frog eat the mealworms when they're still? He only goes for them when they start moving around." A quick answer to such questions can destroy a possible investigation. The children may not realize that they have invitations to new games. The director's role is to help launch them. He can redirect the question: "Have you measured the pendulum at different places? What happened? Suppose you had a bob that was a foot in diameter, where would you measure?" "When a lady wears perfume, can only those standing on one side of her smell it?" "Do you suppose that a frog sees the way we do?" He can suggest some things to do: "Try the softball for a bob instead of the golf ball. See what happens." "Air behaves very much like water in this respect. Why don't you get the task card for making a pressure container and see if it gives you some ideas." "Try feeding the frog by hand and see whether it's a matter of distance, direction, movement, or a combination of all three. By the way, if it's movement, try moving the frog while keeping the food still." Finally, the director can recommend books on the shelf. Whatever he chooses to do, his objective is to get the search going, not to nip it in the bud.

As he continues his rounds, the director may ask a question about the design of an experiment, the records that are being collected, or the form in which the records will be presented. He may recommend a more convenient piece of equipment and instruct in its proper use. He may point out that someone else is

Teaching Science to Children

doing related work and suggest a collaboration. He may ask how much longer it will take to carry out an experiment, not to press for early completion, but to plan for the next presentation. He notes common needs and problems such as measurement inconsistency, poor graphing, lack of adequate controls in experiments, insufficient planning, or failure to observe safety precautions seriously enough. He will raise these problems with the entire group or with that portion which seems to need help.

The director further helps the children by supplying them with whatever equipment they may need. Most games make use of equipment of one sort or another; games in science employ a large variety of useful equipment. There are film loops and projectors, filmstrips and projectors, overhead projectors and transparencies, cases of materials for individual and small group work on many areas of science, models of all kinds, sound films which can be obtained on request, scales, microscopes, various tools, chemical substances, heaters, and a variety of other useful equipment and expendable supplies. While making his rounds, the director notes which, if any, of the variety of resources an investigator or a team may find useful, and he draws their attention to them.

There are some occasions when games stop and attention is given to the director. Perhaps those lovely designs made by Mr. Oglesby with a compound pendulum have arrived. The director holds these esthetically pleasing curves before the entire group and discusses how they were generated; he then hangs them on the wall. He may demonstrate some of the interesting things one can do with soap bubbles and invite someone to add other things they can do. If a snake is a recent arrival, he may demonstrate how to hold the animal so that it will suffer the least discomfort, indirectly encouraging the more timid children to try. There are proper occasions for such things.

This is not traditional instruction where the teacher stands up front and the children sit at assigned desks. Traditional instruction, even under the most favorable circumstances, results at best in the temporary acquisition of information by some children. More often

Games

than not, the factual material is dispensed on a lend-lease basis. "I shall make it available to you today, but you must give it back at the end of the week, month, semester, or during the State examination. Then the slate will be wiped clean, and you may borrow again." The children oblige by keeping the slates as clean as they can. All this has little to do with science, either with learning or teaching it.

EXPERTS CONTRIBUTE

Children can benefit from the special experience of some adults. The laboratory sometimes invites a guest who can share his particular interest. Once it invited an economics professor who brought a number of old American apple-peelers, which he collects. He also brought a big bag of apples. A group had been involved with the ingenious devices used by early Americans to make their labors less burdensome. The children were intrigued, as was the director, by the clever designs of these machines which peeled, cored, and sliced the apples. Of course, everyone wanted to try to do an apple. Here was a new and exciting game.

Another time a remarkably prolific writer of children's science books came with a shopping bag filled with unroasted peanuts, mung beans, and more common seeds. In minutes she infused the children with her passion for growing things. Everyone planted one thing or another, in the process picking up hints about drainage, media for growth, and light. The "lady-from-the-zoo" was always a popular visitor. One never knew what she might bring, perhaps a monkey, a snake, or a hen.

Once a very gifted musician and folklorist demonstrated how to fire and hammer an oil drum to turn it into a musical instrument and how to drill bamboo to convert it into an Israeli-type recorder.

Then there were the "real" scientists. One demonstrated a laser. He showed the concentrated light on the far wall and the paths of light through water as various lens and combinations of lenses

Teaching Science to Children

were placed in the beam's path. Another scientist talked about what he did at his laboratory, how much remains to be discovered, and why time is so interesting. The children were fascinated by the presentations, but they were almost disappointed. These were ordinary-looking men. They might even be somebody's father. One person, who could not leave his work invited us to come to see him. A scholar of the American Indian, he made a fire with bow and drill and demonstrated his technique for chipping flint to make arrowheads. Some of the chilren who had previously struggled heroically with meager results saw that these things *could* be done.

THE LABORATORY TAKES TO THE ROAD

During a trip to the museum it becomes clearer why the nearly symmetrical structures of most vertebrates facilitate the reconstruction of complete skeletons from fragmentary remains. A bone on one side is very likely to be balanced by its mirror image on the other. The children will put that lesson to use on those bones back in the laboratory. There is much more here that the school laboratory cannot provide—that famous collection of insects from one backyard, dioramas of animals in close facsimiles of their natural environment, a profile of the forest floor, a huge mineral collection, and models of the human reproductive process in sequential stages.

The children also visit the planetarium and the zoo, other places where the things they want to see have been carefully collected, organized, and displayed for their pleasure and enlightenment.

They also like to see the movement and coordination of work, the use of materials and machines, the lively world where "it's happening." Ships loading and unloading; buildings rising; machine shops where thousandths of an inch is a common measurement and where the spinning grinding wheel appears at rest; the smelly, rowdy fish market; the aromatic, warm bakery; the clean,

Games

quiet nuclear plant—all these and more are accustomed to their visits.

At the heart of all these activities is the game. Wherever the game leads, to the heart of the city, to the countryside, or to the base laboratory, there is time to play, to explore, to manipulate, to reflect, to make wild leaps of the imagination. The rhythms of the play invite children to invest head, heart, and hand with the assurance that the search, not the bell, is the arbiter of time, that each player will be heard, regarded, and criticized, not measured, marked and filed.

These are games in which even the errors are interesting, and recitation of the verities may be greeted by doubt. Thought and feeling live in happy symbiosis. Ennui, the inevitable consequence of the denial of the wonder of both science and children need not be overcome because it has never been allowed to enter.

4. Questions and Problems

> Hast thou entered into the springs of the sea? or hast thou walked in the search of the depth? . . .
> Hast thou perceived the breadth of the earth? declare if thou knowest it all.
> Where is the way where light dwelleth? and as for darkness, where is the place thereof. . . .
> Hast thou entered into the treasures of the snow? or hast thou seen the treasures of the hail. . . .
> By what way is the light parted, which scattereth the east wind upon the earth?
> Who hath divided the water course for the overflowing of waters, or a way for the lightining or thunder;
> To cause it to rain on the earth, where no man is; on the wilderness, wherein there is no man;
> To satisfy the desolate and waste ground; and to cause the bud of the tender herb to spring forth?
> Hath the rain a father? or who hath begotten the drops of dew?
>
> Job 38:16, 18-19, 22, 24-28

In the last chapter it was recommended that children be left to develop their own tactics in the game of investigation. To do that they must first have a question or problem sufficiently interesting to elicit their maximum energy in tracking down an answer. Finding interesting questions and problems, the single most imaginative and demanding step in scientific investigation, is not served well by the logical mood. That mood is useful for verification. To imagine the question or problem, the playful, intuitive mood of an interested, informed, and autonomous mind is needed. Try to pose several questions that you consider interesting, significant, or both,

Questions and Problems

to which you presently do not have any answers. You will agree, I believe, that questions of worth cannot be compelled or scheduled. They are the leap of an imagination steeped in rich, absorbing experience. Accounts of scientific discovery attest to the creative character of finding questions.

Ilya Mechnikov, a Russian bacteriologist who received the Nobel Prize in 1908, was absorbed by the nature of intracellular digestion in protozoans. *The Art of Scientific Investigation* (1957) by W.I.B. Beveridge contains Mechnikov's story of a crucial episode in his investigations: "One day when the whole family had gone to the circus to see some extraordinary apes, I remained alone with my microscope, observing the life in the mobile cells of a transparent starfish larva, when a new thought suddenly flashed across my brain. It struck me that similar cells might serve in the defence of the organism against intruders. Feeling that there was in this something of surpassing interest, I felt so excited that I began striding up and down the room and even went to the seashore to collect my thoughts." The question whether the biological phenomenon observed in lower forms might be operative in higher forms was in no way routine. It could not be anticipated. According to Mechnikov, "It is easy to prove that amoeboid cells are capable of enveloping and digesting (more or less completely) all extraneous matter" (*Ilya Mechnikov, His Life and Work* [1959] by Semyon Zalkind). Nowhere did Mechnikov write that it was easy to formulate the *question* about phagocytosis, the process by which certain cells in the blood, phagocytes, engulf into their cytoplasm particles from their surroundings and thereby help protect the body against disease.

Indeed, it is not easy to ask such questions. For centuries scientists debated whether life is generated spontaneously under ordinary conditions. But it was left to Louis Pasteur to ask: "Are there germs in the air? Are they present in sufficiently great numbers to explain the appearance of organized productions in infusions which have previously been heated? Is it possible to obtain an approximate idea of the relationship between a given volume of

Teaching Science to Children

ordinary air and the number of germs which this volume contains? [If] . . . it can be recognized that there are constantly in the ordinary air a variable number of corpuscles . . . Are there really fecund germs among them?" In the remarkable monograph from which these questions were extracted, *Memoir on the Organized Corpuscles which Exist in the Atmosphere, Examination of the Doctrine of Spontaneous Generation* (1862), Pasteur remarks, "The procedure which I followed in order to collect the dust suspension in the air and to examine it with a microscope is very simple." It is unquestionably true that techniques which are very simple for Louis Pasteur and Ilya Mechnikov are not at all simple for most people. But it is far more difficult to think of their questions. René Dubos in *Pasteur and Modern Science* (1960) informs us that Louis Pasteur "was a masterful technician but also highly intuitive." We can strive to imitate Pasteur's technical mastery, but each individual must cultivate his own intuition.

The intuitive leap, the choice of sequence among problems, does not take place in a vacuum. Each scientist's creative imagination is fed by the various elements in his social and cultural environment to which he responds according to his unique interests and gifts. The scientist is both a product of his time and a progenitor of a new time—the cultural context allows for his development, his uniqueness senses the novel and formulates the questions that point ahead.

Before the H.M.S. *Beagle* set out on a voyage of scientific exploration with Darwin as its naturalist, Lyell had already advanced the idea that present observations of geological forces and conditions can explain geological formation. Smith had shown that certain strata contain particular kinds of fossils. By the early nineteenth century Jean Baptiste Pierre Antoine de Monet Lamarck, the French naturalist, had established the unity of life, the capacity of species to vary, and had emphasized the importance of environmental influences. In *Darwin's Century* (1961) Loren Eiseley tells us that "The gardens and paddocks of kings and

Questions and Problems

nobles were revealing what curious, never-before-seen varieties, historic shapes in other words, could be turned from the darkness of non-being by the selective hand of the breeder." All of these, and many, many more, were elements in the upwelling of the nineteenth century. But it was Charles Darwin who asked in *The Origin of Species:* "Can the principle of selection, which we have seen so potent in the hands of man, apply under nature?"

ONE QUESTION LEADS TO ANOTHER

Darwin answered that question magnificently, but he was not able to solve the puzzle of the means by which heritable characteristics of the species are realized in subsequent generations. In the late 1860s, almost ten years after the publication of *The Origin of the Species*, he was still elaborating his provisional hypothesis of pangenesis according to which the transmission of gemmules which represent various organs and components of the body accounts for the observed similarity in characteristics from one generation to the next. Some twenty years later, in his essay "The Continuity of the Germ-plasm as the Foundation of a Theory of Heredity," August Friedrich Leopold Weismann, a German biologist, asked "How is it that a single cell of the body can contain within itself all the hereditary tendencies of the whole organism?" His pursuit of the question with experiments in which he cut off the tails of mice for several generations led him to the germ plasm theory. Although the theory is inadequate because it does not hold for all higher animals and holds not at all for lower animals and plants, Weismann's question led him to make important contributions to our understanding of the chromosomal mechanism of inheritance.

Good questions are followed by investigations which attempt answers. Answers inevitably raise new questions. E.E. Cummings acknowledged this in the introduction to his *Collected Poems* (1923): "Always the beautiful answer who asks a more beautiful question."

Can mathematics, through methods for analyzing frequency distributions, contribute to a mathematical theory of evolution? How many pairs of chromosomes are there in man? Are the structure of cells and heredity related? What is the size of genes? Can mutations of the gene be induced artificially? Can genetic factors such as selection pressure, mutations rates, and isolation help to organize our understanding of evolution? What is the structure of the gene? Is an understanding of the chemical structure and dynamics of the gene sufficient to explain heredity? To know more is only to realize how much more there is to know. That is one reason that the rate of growth of scientific achievement accelerates. Almost every ten years, it has been estimated, the amount of scientific information doubles. And it all begins with questions.

"TO QUESTION" IS A VERB

If children are to learn science, they need an environment which makes asking questions a comfortable thing to do. Very young children, those who have not yet entered school and those in the primary grades, are filled with questions. They ask very hard ones, such as: Why do people die? Is it a kind of sickness? What made the sky? What made the sun? Who makes bugs? Where do stars come from? They are particularly interested in origins. In a delightful book about the language of young children, *From Two to Five* (1966), Kornei Chukovsky, a favorite author among Russian children, records the following question by a child: "Mothers give birth to boys too? Then what are fathers for?" Children's questions deserve respectful attention. They cannot be dismissed if children are to learn from their earliest recollections that questions are the entry to mystery, that their interest is shared by adults, that ignorance is only in degree and is always present in small or large

Questions and Problems

measure, and that to reduce it is one of the great pleasures of life. We can all share the feeling of the young man whom George Eliot describes in the introduction to *Romola*, "the night-student, who had been questioning the stars or the sages or his own soul, for that hidden knowledge which would break through the barrier of man's brief life, and show its dark path, that seemed to bend nowhither, to be an arc in an immeasurable circle of light and glory."

It is illuminating for children to discover that what they had thought were their own private interests and concerns are shared by a vast number of people living in many different places and at different times. Legends convey this feeling very well. They are man's storehouse of his questions, hopes, and fears, inscribed by his poetic imagination. They were fashioned from his observations, and they were conditioned by the social life with which he was familiar. In *Introducing India* (1967), Raj Thapar tells the story of Vishnu. "Vishnu, who preserves all life, dreamed of a universe. A beautiful golden lotus sprang from his navel. It was a flower of wonder with a thousand petals. The light which came from it was the light of the sun. From this first lotus was born the god Brahma who created the universe. Our earth sprang to life as if out of a dream, out of a flower." This is a metaphorical expression of the story that George Wald, the Harvard biologist who received the Nobel Prize for medicine and physiology in 1967, tells his students in the introductory biology class. In his story, George Wald traces the origin of living forms on earth to the formation of chemical elements in the stars. We are, indeed, such stuff as stars are made of.

In the story the Cherokee Indians tell of the creation of human races, God baked the races; the white race was underdone. Children listen to the *Tales of an Ashanti Father* (1967) as told by Peggy Appiah. They hear how death came to mankind because of Kwaku Ananse's greed, one of many *anansesem*, or spider stories, from the rich Ashanti tradition. The Crow Indians have their Old Man Coyote who molds the earth, creates mankind, bids them multiply, and instructs them. The Dobuan islanders of Papua tell us how fire was created. The Mayan villagers of Yucatan have their Itza who lived during the good times, another version of the loss of paradise.

Children love these stories. They listen intently, and when the stories are over, they challenge the adequacy of the explanations with countless questions and with evidence in rebuttal. But if you listen while a child who is familiar with these stories tries to explain some of the cosmic mysteries to a child who has not heard them, you will often hear the stories repeated. The stories affirm that the questions which absorb young children have a long and honorable history among all mankind.

Having established the propriety of asking questions in the minds of the very young, and having demonstrated the interesting variety of forms questions have taken among people in the most diverse circumstances, we can consider questions differently. Assisted by the imagination and ingenuity of those who searched the highest heavens, the depths of the earth, and all between, we can share with working scientists the wonders of the very small and the inconceivably large.

This is a good time to introduce children to the many excellent biographies of scientists that have been written for children. They can read about Galileo, Newton, Priestley, Pasteur, and Einstein. The children who cannot yet read independently can join with accomplished readers from their own age group or with older children and listen to the accounts they read aloud. Think of the possibilities for significant grouping, for real social interaction, as these rich mixes hear the stories of new heroes, stories eminently suited to children who have seen men jumping on the moon and exploring the depths of the ocean.

Biography in science illustrates that those who found solutions to problems did so because perplexing questions gave them no rest. Questions were both intellectual and esthetic high moments in their lives. We must strive to make them so for children. Questions should be greeted with the enthusiasm they deserve.

Questions and Problems

The mother of James Clerk Maxwell recorded that even before he was three years old, the words " 'Show me how it does' [were] never out of his mouth." He also constantly asked, "What's the go of that?" Maxwell's childhood and those of two other scientists are very well described in *Faraday, Maxwell, and Kelvin* (1964) a delightful little book by D.K.C. Macdonald. *Crucibles* (1939) by Bernard Jaffe reveals that Irving Langmuir, the American scientist who won the 1932 Nobel prize in chemistry, asked while he was still a young child, "Why does water turn to ice? Why does water boil in a kettle? Why does rain fall?" When the mature Robert Hooke wonders why cork is so very light, there is a childlike quality in the question. The same freshness, daring, openness, and naïveté unburdened by need for certainty characterize the questions asked by young children. Something of this quality remains with those who become innovators in science, just as there is something childlike in the rapt concentration of gifted poets.

If the child's gift of questioning is not nurtured, it is soon lost. A study by J. Richard Suchman, "Inquiry Training: Building Skills for Autonomous Discovery," confirms this. "We got the feeling that many of the children saw no connection between asking questions and discovering causal relationships. It was clear that they had not had much experience in asking questions and did not feel comfortable in their new role. The main problem seemed to be the strangeness of the task. The rewards in the classroom had typically come for giving the right answer." Just a few years in school and questions have become strange and unrewarding. Most children fall victim to the answer syndrome and to the periodic "normalized" examinations that purge thought and criticism (and an occasional breakfast). Few children can resist the answer syndrome, and once they have succumbed to it they find it difficult to believe that intelligence is not given by facile answers, just as few of their parents can resist the cash syndrome, and find it difficult to believe that Socrates was right, "that virtue is not given by money." There are some hardy souls who withstand the onslaught of demands for superficiality and the entrenched institutional fancy that glibness and passivity provide passage from ignorance to wisdom. Poor Socrates! He might never have made it past the first round of a quiz program.

KEEPING THE SPIRIT ALIVE

And yet, the delight in questioning can be sustained. What is needed is a learning environment which encourages rather than discourages inquiry. A group of children engaged in an activity devised by the imaginative workers at Elementary Science Study in Newton, Massachusetts, insert a thermometer in a paper cup filled with water. The cup of water with the thermometer inside is placed in the freezer. When the water has frozen solid, and the paper cup is peeled off, what is left is a kind of popsicle, the thermometer serving in lieu of a stick. The ice is held over a candle flame, and the reading on the thermometer is observed. As the heat of the candle flame melts the ice, the reading on the thermometer shows no change. Can you imagine that children who have had experience with a thermometer will *not* ask questions?

Another group has inflated a balloon and tied it off. They have been invited to push against the balloon with their hands. Some children push quite hard and alter the balloon's shape. However, when they are given a pin with which to touch the balloon, it pops even though they apply very little pressure. Is the problem resolved by calling the pin "sharp"?

The landing on the moon prompts a discussion of the exquisite engineering displayed by the astronauts' equipment. It has been estimated that there were some 17 million parts in the machinery that took the astronauts to the moon and brought them back again. If the machinery is granted a reliability of 99.9 percent, a very generous allowance, it is still amazing that they succeeded. A little arithmetic is required to recognize the problem.

The same group analyzed the news media reports of the moon landing. Included in some reports was illiterate chatter about how

Questions and Problems

the weightless condition of the astronauts was the consequence of their having gone beyond the effect of the earth's gravity. Free fall went unmentioned. Then the children considered some features about the flight with which they were already familiar. They knew that the earth dips some sixteen feet every five miles, that if they watched a boat measuring sixteen feet from the water line to the tip of the mast sail away, the last thing they would see would be the tip of the mast, and that would disappear when the boat had traveled five miles out to sea. They knew that if it is true that a freely falling body falls sixteen feet during the first second of fall, then a satellite must travel at about five miles a second to stay in orbit around the earth. In one second it will fall sixteen feet, but by that time it will have gone five miles, and the earth will also have dipped sixteen feet, and the satellite will be just as far from the earth as it was a second before. Five miles a second is 18,000 miles an hour. The satellite is traveling at this tremendous speed parallel to the earth's surface. Why does it not just go off into space? What keeps it hanging around? Furthermore, in order to plan the trip to the moon, the moon's position had to be predicted very accurately. How is this done? Such exercises in event analysis invite questions and serve well for political, as well as scientific, news. Suppose that the astronauts had discovered a mosquito aboard their craft. Can a weightless mosquito sting?

There are many ways to keep questioning alive. One technique is to demonstrate without making any comments at all. Two hacksaw blades are fastened to the end of a table with C-clamps. One blade is fastened near its end so that almost its entire length extends beyond the table. When it is struck to start it vibrating, the only sound heard will be that of the blade striking against the table. There may be no sound at all. The other blade is fastened so that only about three inches of it extend beyond the table. When this one is struck, there is a clear tone.

A pint of strawberries is displayed. The impressively large ones are on top, just as they usually are in the store. The berries are removed from their container and placed in another box of similar size. A lid is placed over the box of berries, and the box is shaken vigorously. When the lid is removed, the larger strawberries are again on top.

During school holidays there is always the problem of providing water for the plants. One way to supply them with water in the absence of their normal caretakers is to let the plants "drink" steadily but slowly. A container is found for each potted plant that is large enough for the pot to enter, but not sufficiently large to permit the pot to slide all the way to the bottom. One end of a stringlike length of twisted absorbent cotton is inserted in the hole at the bottom of the pot. The other end is allowed to rest in the supporting container which has been filled with water to within a few inches of the pot bottom. The water from the container climbs up the cotton into the soil of the potted plant.

Everyone has seen a statue of a horse with one leg off the ground. Our "horse" is a table with folding legs. One leg is folded. The table is standing on three legs. On top of the table and to one side is a box containing scale weights. Under the table, directly beneath the box of weights, a plumb line and bob are suspended. On the floor a triangle has been drawn whose corners are the three standing legs.

Earthworms are cut behind the clitellum, the saddle-like region, into two halves. The cut worms are left on moist soil in the terrarium. After several hours half of the parts have disappeared.

The "equalizer machine" makes the littlest girl in the group a match for the two strongest boys in a tug-of-war.

Sometimes the demonstrations serve a double purpose. They stimulate questions, and they may also serve as opportunities to introduce a piece of equipment that the children have not yet used. While the children watch the screen, the director places a right circular cone, a circular cylinder, and a sphere (resting on a bit of clay to prevent rolling) on an overhead projector and turns the projector light on. The shadows on the screen all look alike.

The hand-operated planetarium is demonstrated. It would ap-

Questions and Problems

pear from the model that we should have twelve solar and twelve lunar eclipses a year.

The children can always find something of interest in the puzzle corner. There is a target that "refuses" to accept a bull's-eye landing by a craft lowered to its surface. At the very last moment, the craft veers to one side or another. There is a jar with masked coins inside. The magnet that is tied to the jar attracts only one of the coins. A can, when placed on its side, rolls in most positions. Occasionally, it does not roll. A homemade spinner consists of a disc of white, glossy cardboard pierced through the center with a dowel which has been sharpened at one end like a pencil. When several drops of ink are placed on the upper surface of the disc, and it is spun, the ink drops make tracings. A whiskey shot glass filled with oil stands inside a water glass containing just enough alcohol to cover the whiskey glass. With the help of a medicine dropper, water is allowed to drip slowly down the side of the water glass. Suddenly, a sphere of oil, or sometimes a "flying saucer" rises out of the whiskey glass and remains suspended. If a wire is carefully lowered into the suspended oil and spun between the palms of the hands, the oil figure spins too, and while it spins it becomes flatter, finally sending off little drops from its sides. There is the ever popular Cartesian diver, the cork fisherman who teeters on the edge of the table but does not fall, and many more puzzles that appeal to varied interests and tastes.

The bulletin board, too, is used to stimulate questions. Sometimes it announces a bit of straight news—"A radar signal sent from the earth and reflected from the moon's surface took two and a half seconds for the round trip." Or it may refer to one of the enigmas that many people consider fleetingly without realizing that the same thing has puzzled some of the greatest minds. One such enigma is simply "The sky is dark at night" (Olber's paradox). Another one is "The moon, which is smaller than the earth, covers the entire sun during a total eclipse of the sun." Another entry contains a bit of a clue. "The speed of a running animal depends upon the product of the length of its leg and the frequency with

Teaching Science to Children

which it can be swung. Consider your work with the pendulum." The effort to stimulate questions takes a variety of forms, but there is no attempt at exercises to "find the question." Such exercises are simply a new disguise for "What's the answer?" except that the question has become the answer. The efforts have served their purpose if they generate in some children a desire to push on, to find "how it does" or "what's the go of that?" in which questions are always implicit. When questions are actually posed, the children consider them. Sometimes they launch an investigation.

5. Head and Hand

> If one doesn't enter the tiger's den, one cannot obtain tiger cubs.
>
> Old Chinese saying

When children actively search their environment—explore, play games, and question—they are engaged in perceiving segments of their world. Perceiving is a necessary step toward knowing, but it is not the only step. To know the environment it is also necessary to come to grips with it in a practical way. A child who has learned to tell the time from a clock has gained a technical skill. Knowledge of the meaning in the clock's message becomes more profound after he has observed the return of a star to the same point in the sky some four minutes earlier on each of successive nights and then calculated how long it will be before the star will return to the point in the sky where it was first observed. And in order to follow the star, he must construct a tracking device, however primitive. To know in this deeper sense, to combine the imagination of the mind with the cleverness of the hands, is a distinctly human characteristic, one inherited from man's bio-cultural tradition. It is a characteristic all infants announce in that superb moment when they first discover their hands. The meaning of that dramatic gesture proclaims their uniqueness in the animal kingdom.

It has become fashionable in recent years for professional and amateur animal ethologists to point to the many interesting ways in which human beings resemble their animal cousins, including some very distant ones. Geneticists, ecologists, ethnologists,

Teaching Science to Children

molecular biologists, and other scientists devoted to the study of life have many interesting and significant things to teach about similarities among animals but they are also concerned with the differences.

When they examine a species, among the things they want to know are the characteristics specific to it. For example, the members of the trypanosome family are all one-celled, parasitic animals. However, while the species *lewisi* usually does no harm to its host, the species *gambiensi* and *rhodesiense* are responsible for two kinds of human sleeping sickness in Africa. Scientists want to know not only how animals are related, but the demarcation that establishes their uniqueness. The species-characteristic that most clearly defines man is culture.

HUMAN UNIQUENESS

A mark of humanity is the ability to learn and to transmit what is learned to children. Without this capacity it is questionable whether man would be here at all. His sensory acuity is no match for that of many animals. He cannot compete with the hawk or the vulture in seeing distant objects, nor with the little tarsier in seeing at night. Man's range of hearing is less than a dog's, and his sense of smell is not as discerning as the dog's. Think of the local news a dog can gather with just a turn or two around the corner hydrant. The catfish and carp have their taste receptors distributed over their bodies, and the common fly can taste with its feet.

Some animals possess senses for which there are no human counterparts. Whales and bats emit calls whose echoes inform them about their environment. Some fishes can determine whether there is dinner in the neighborhood by emitting electrical discharges. Pit vipers count on their ability to detect faint differences in heat to find their meal. Bees can polarize light. Birds and some

Head and Hand

turtles find their way about the world by means which are not understood.

And yet, despite his limitations, it is man who has found the way to see the craters of the moon and the moons of Jupiter. He can sense the arrival of an echo from Mercury and Venus, and with the information, he can deduce their rotation rates. Man has produced Polaroid material, he has built instruments to guide him about the world and he can teach his children how to use them.

The natural strength of humans is not superior. No man can match the ant's feat of supporting a thousand times its own weight. Nor is man as fast as the cheetah, which has been clocked over short distances at 71 miles per hour. When human strength and speed prove insufficient, man's nails will not protect him as well as a racoon's claws protect the raccoon, and if he crouches with his hands over his head, the protection they provide is no match for that of an armadillo's armor. Yet, the ant's remarkable strength is dwarfed by the operator of a shipyard crane. Even if he is a heavy man, he can lift well over 3,500 times his own weight. No known animal is fast enough to escape from earth's gravity, but man has found a way, and he has been so successful in protecting himself that he is now concerned about a population explosion. However, serious questions do exist about how well he is safeguarding the continued existence of the human species.

Clearly, man's ability to receive more varied and precise information about the conditions of his environment is not the consequence of his anatomy. Nor do the homeostatic adjustments he makes so that his life may go on in the face of considerable environmental variation depend exclusively on his biological endowments. The Lapps in arctic Finland and the Italians in Sicily are not significantly different biologically. Rather, man's ability to extract information from a wide variety of environments and to reflect upon what he has found facilitates his survival in those environments. This ability is the result of man's use of his brain as a coordinating organ to invent social organizational patterns of life,

Teaching Science to Children

and to synchronize the use of his hands with his vision. The social invention and the coordination of eye and hand stimulated what C.H. Waddington, the distinguished English biologist, calls man's "exploitive system" and made us human.

The social-organizational invention in the beginning of man's development utilized his biological capacity to learn and produced culture, man's most distinctive achievement. All that man is and all that he knows are the products of both his biological and cultural existence. Inventiveness enabled man to overcome his biological limitations.

Human beings can invent organizational forms and technical means to adapt to the contemporary natural and social environment and to assure that man will continue to flourish. To make this adaptation successfully requires both the understanding that it is necessary and possible and the desire to accomplish it. Children's science education, if it is worthy of them, contributes to their understanding of the necessity by enabling them to comprehend the interdependence which exists among all living things and between life and the inorganic environment. Such an education incorporates the humane values inherent within scientific activity, making them the criteria for admirable conduct, and contributes to children's desire for mankind's well-being and happiness. Significant science education reveals to children the power of their own inventiveness and thereby contributes to a conviction that there are no necessary and desirable tasks that mankind cannot complete.

THEORY AND PRACTICE

Inventiveness is a human characteristic because men are, indeed, as Prospero taught us, "such stuff as dreams are made on." Other animals may dream, but man alone works to make dreams come to life. Men do build in air, not only trivial castles, but more significant structures. And while these insubstantial figments hang suspended, men lay foundations for their realization in life. Mind and hand, theory combined with practice, place human beings in a unique relationship with their environment. The combination of thought with manual work allows them not only to sense and understand their environment, but to consciously change it to meet their needs.

Despite the interdependence of thought and manual work, those concerned with children's education tend to emphasize the former and neglect, if not denigrate, the latter. Thought is admired and respected as deserving of disciplined study, but manual effort is frequently relegated to a peripheral hobby status.

This traditional separation of the intellectual and the manual can be traced back to ancient times. In the opinion of Andreas Vesalius, the great Flemish anatomist of the sixteenth century, the sciences, especially the medical sciences, went to ruin in Rome after the barbarian invasions because "the more fashionable doctors, in imitation of the old Romans, began to despise the work of the hand. They delegated to slaves the manual attentions they judged needful for their patients, and themselves merely stood over them like master-builders." In *Head and Hand in Ancient Greece* (1947), Benjamin Farrington, Professor of Classics at Swansea University in Wales, has translated Vesalius' elaborate Latin into English. "When the whole conduct of manual operations was entrusted to barbers, not only did the doctors lose the true knowledge of the viscera but the practice of dissection soon died out, doubtless for the reason that the doctors did not attempt to operate, while those to whom the manual skill was resigned were too ignorant to read the writings of the teachers of anatomy." The attitude of these doctors that some men are fit to think while others are fit to use their hands, that doctors think and barbers dissect, was not of their invention. It reflects their social structure. Farrington informs us that according to Xenophon, Socrates believed that "What are called the mechanical arts carry a stigma and are rightly dishonoured in our cities." Such thinking ignored the significance of the Greek word *cheirourgia*, which gives us our word "surgery." It means manual operation. Whatever opinions about theory and practice the philosophers and their colleagues in the literary culture may have held, they wanted to be helped when they were ill or dying and they needed effective doctors to provide that help. From Alcmaeon in the fifth century B.C. to Galen in the second century A.D., doctors retained a privileged position. Despite the general disrespect in which manual work was held, the doctor, who remained a manual worker, was respected.

The scientific experiment, above all other kinds of activity, requires the unity of theory and practice if success is to be achieved. Experiments are likely to be more successful if both a clever mind and clever hands are employed. Most people are more clever with one or the other, few are equally talented with both.

In 1857 Michael Faraday, the English physicist and chemist, wrote a letter to Maxwell in which he asked him to try to write his conclusions in common language instead of in mathematical formulas. Faraday was no mathematician. One of ten children in a blacksmith's family, Faraday went to work at the age of twelve and did not hear his first science lecture until he was twenty. He never mastered mathematics, and it is sheer chance and our good fortune that his remarkable scientific gifts were not lost to the world. Faraday's mathematics was weak, but he was very imaginative and wonderfully clever in the laboratory. Thanks to those gifts the average American family benefits from the silent energy equivalent of thirty-seven workers. Each time someone turns on an electric light, toasts bread, or employs any of the numerous electrical devices that have become commonplace in American life, he is utilizing another of the invisible workers made available by Faraday.

I cannot think of another scientist who succeeded in contributing so magnificently to science with so little knowledge of mathematics, but Faraday proved that it is possible. There are places in science for people with different strengths. In fact, different kinds of experiments may call for different kinds of ability.

EXPERIMENTS AND EXPERIENCE

In one type of experiment things are made to happen so that the scientist can observe—as when Galileo went home and set up a pendulum. This is the type of experiment that children do most frequently. Another common kind of experiment is one in which the scientist makes things happen to check up on his ideas. Pasteur's famous experiment on microbes is of this variety. Sometimes one scientist has the idea and another scientist does the checking. Einstein had an idea about the relationship between light and a gravitational field, and Sir Arthur Stanley Eddington, the English astronomer and physicist, checked up on it by observing the total eclipse in 1919. Eddington's observation helped to establish the validity of Einstein's idea.

Not only are there different kinds of experiments, but there are also many ways to do an experiment. There really are no rules except honesty, great care, and persistence. There is no known formula to describe scientific methodology.

Unfortunately, scientific papers do not include the numerous little interactions and the flavor of events in the laboratory. In *The Art of the Soluble* (1967), P.B. Medawar suggests that listening at the keyhole would be very helpful in gaining a sense of the laboratory atmosphere. He imagines that one might hear something like this:

"What gave you the idea of trying . . .?"

"I'm taking the view that the underlying mechanism is . . ."

"Actually, your results can be accounted for on a quite different hypothesis."

"It follows from what you are saying that if . . . , then. . . ."

"Is that actually the case?"

"That's a good question" (that is, a question about true weakness, insufficiency, or ambiguity.)

"That result squared with my hypothesis."

"So obviously that idea was out."

"At the moment I don't see any way of eliminating that possibility."

"My results don't make a story yet."

"I'm still at the stage of trying to find out if there is anything to be explained."

"Obviously a great deal more work has got to be done before. . . ."

"I don't seem to be getting anywhere."

One can imagine the involvement, the uncertainty, and the free exchange of questions and criticism in a laboratory where experiments are in progress. If children can be helped to understand that their experiments may take a variety of forms, that their modest efforts serve well to the extent that they reduce ambiguity and are reported clearly and economically, then their experiments are likely to be personally satisfying and educationally significant.

The children's involvement begins with the design of the experiment—itself a creative piece of work. The doubts, the questions, the problems—the interest—are theirs, and each child must design his own experiments if he is to satisfy his interest.

Of course, children rarely think of questions that have not been asked before and so it is probable that some instructions and possibly a diagram will be found in a book, particularly in the physical sciences. If the book satisfies, that is all right. Living forms and their habitats are so many and so varied, their structures and functions so complex, that children's questions about the biological sciences have not necessarily been answered—"What makes something taste sweet?" "When I cut myself, new skin grows. How does the skin 'know' when to stop growing?" "How do birds find their way over the ocean?"

Whether the question is original or not, it is more likely that attempts will be made to design an experiment if it is provoked by

Head and Hand

something available to the child. If some children choose to make a "search of the literature" for more information or for ideas, this, too, is part of the game. Some will struggle a while and then settle for very little while others will follow the chase until they are fully satisfied. One child may quit after growing a little mold and reading some textbook information about it. Another child may not rest until she has tried every conceivable environment in school and at home in which mold can and cannot be grown, has tested her observations, and has drawn some general conclusions. A third child may be entirely pleased by the observation of the colors a prism produces in the sunlight. Careful observation is a kind of experiment too. It is, after all, what classifiers and many ecologists do.

Sometimes a child struggles to simulate an event he cannot reproduce directly. It may be that he has followed what he has read about refraction, but he would like to see the process as well as the result. He thinks of an experiment that will reproduce the conditions of refraction, but by analogy. He lays a blanket across half of the top of an inclined table. He lets a spool from camera film roll down the table. He traces the spool's path when it is at right angles to the boundary between the edge of the blanket and the table, and when it is not. He may even try to predict under which conditions, table to blanket or blanket to table, the path will be nearer to the "normal." If he can devise some clever way to measure the different velocities on the table and on the blanket, he will be able to calculate the "refractive index of the blanket" by finding the ratio of the two velocities. Whether one is a Marxist or not, the slogan raised by Karl Marx in the *Critique of the Gotha Program* makes wonderful sense in teaching children: "From each according to his capacity, to each according to his need."

Children's need to experiment cannot be met without a great variety of materials that invite children to try to make something happen, to impose change and observe the consequences. Materials may be found about the house—milk cartons, old electrical appliances, spools, combs, switches, old vacuum cleaners, mechanical toys, bottles, jars, etc. Or they may be obtained from supply houses and local stores—scales, hot plates, thermometers, balloons, live animals and plants, dry cells, wire, bulbs, magnets, and many more necessary commercial articles. There are also collections of materials developed by various research programs in elementary science. If the circumstances are encouraging, the right materials wear invisible signs that say "CHANGE ME" to children just as clearly as the cake said "EAT ME" to Alice.

The changes the children make will hardly be credited with the status of experiment by strict constructionists. Children who boil milk, assemble a chicken skeleton, or wire a circuit just to see what happens are having what Medawar says the French call an *expérience*. But the name does not matter. What matters is that the children work to some end that makes sense to them, that they express their results in a clear and concise manner, and that whatever explanations they offer flow from the evidence. Some people become very uneasy about the children's unsophisticated and unpretentious explanations. It is helpful to remember that all scientific explanations are approximations, and that the explanations offered by the children reach a very low level of approximation. They are, after all, very young, and one ought not to expect more. It is more exciting, and more significant too, for children to combine observations in some way that makes new sense than it is for them to receive a magic box containing someone else's answers which may make very little sense. Making new sense is what the experiment is all about as far as the children are concerned. They searched and they found something interesting.

What they have found may be very modest by professional standards, but earth-shattering from the children's point of view. The realization that more turns of wire about a nail makes a more powerful electromagnet can be very exciting to a child. If he is told that Joseph Henry made the very same discovery one hundred and forty years ago, the child is not disappointed. He is more likely to think that Henry was a rather clever fellow. The news in no way diminishes the frontier quality of his effort. He feels more kinship with Francis Bitter, the American physicist at M.I.T. in the 1930s, than with Henry, and follows up the initial investigation with an examination of as many variables as he can imagine. Is there magnetism without the iron nail? More? Less? What happens if the wire is not insulated? Suppose two batteries in series are used instead of one battery? Can the poles be anticipated? How is the strength of an electromagnet measured? What happens if the nail is bent into various shapes? The child cannot yet understand Feynman's explanations, but he is engaged in acquiring the raw experience, the substrate to which explanations will respond, the quality of the response in large measure governed by the quality and extent of these early experiences. If we sustain the child's interest, we increase the likelihood that one day he will be able to read *The Feynman Lectures on Physics* (1964) intelligently, whatever his professional career may be.

Sometimes an experiment seems to lead only to dead ends. No one can promise that effort, however devoted, will result in success. Recall the story of the little boy and the ceiling tiles at the beginning of the book. His plan for a sound accumulator was defeated, but he was not. His restless dissatisfaction with the strange turn of events prompted him to find out whatever he could about sound. In his search he came across the suggestion that sound generated by a tuning fork could, under certain conditions, blow out a candle flame. What had appeared to be the end of the road turned out to be a traffic circle with a number of new roads, any one of which he might choose to follow.

Commonplace articles often extend invitations to follow. Unframed plane mirrors lure children to investigate—they can inspire a variety of experiments. The discovery that when the mirror is stood at right angles to the table across the middle of the word "ICE," the word is reconstructed, suggests the possibility of new experiments. There is considerable satisfaction when "WAIT" is lettered vertically, and the mirror is stood on edge vertically across the middle of the word. New combinations of letters are tried, but no longer arbitrarily. Some try to hold objects with one axis of symmetry before the mirror, then those with more than one, and still others that are asymmetric. The atmosphere grows livelier

Teaching Science to Children

when someone tapes two mirrors along one edge. Strange effects are noted as the mirrors are turned. At ninety degrees an argument begins.

"Look, you're turned the other way."

"No, I'm not. When I blink my right eye, the right eye in the mirror blinks."

"But it's not on the same side. Your right is here (pointing to her eye), but it's over there (pointing in the mirror) in the mirror."

"But in the mirror I'm facing toward me."

The first child cannot follow the argument and remains silent.

"Pretend that you're in the mirror. Stand here facing me. In this mirror (two mirrors at right angles) if I blink my right eye, which one do you blink?"

The first child blinks her right eye and smiles.

"You see, if that was me in the mirror, it would be my right eye."

"Let's try just the one mirror again," the first child suggests.

With these modest beginnings we introduce children to a wider and more significant world. They will now perceive the building and the airplane, the snowflake and the painting in a new, more responsive way. If we build on these early experiences, the symmetry of an arrangement and the idea of a "group" are just small steps away. The children who have acquired these rich experiences are more likely to find pleasurable and relevant a consideration of degrees of order and disorder in the portions of our natural environment that attract them, in atoms, crystals, the living cell, the earth on which we live, or the sky above.

Other children are embarked upon different experiments.

"The earth is a magnet, isn't it?"

"Yeah, but it's weak."

"Strong enough to pull the compass needle."

"Which way will you hold it?"

"In a north-south direction."

One child holds a long rod of soft iron across his shoulder. Problems arise about testing for the induced magnetism they anticipate. If magnetism is detected, can the rod be de-magnetized by holding it in an east-west direction? How long will the magnetic

Head and Hand

effect last? Is it possible that some things in the room are already magnetized? What things should be tested? How? Will the orientation of things in the room make a difference? Will all parts of the same object—say a radiator—respond similarly?

In the corner, a girl experiments on her own, filling spaces with geometric figures. She wants to know whether bees have hit upon something unique in the construction of their hives. The experience of seeing a living hive through a window has not left her. She was intrigued by the elegance in the construction of the cells.

Several children have before them garnet and quartz crystals, a sheet of mica, several pieces of obsidian, some flint, table salt and several other common salts, and alum. Should an adult direct their activities toward a particular discovery? But who is to say what they "should" discover. There are volumes on salt. They certainly cannot be expected to discover all that. What, and how much, will they learn?

On first examining the quartz and garnet crystals, the children are struck by their beauty. They wonder about how they are put together. Perhaps it will be enough for some to discover that the crystals of table salt from the box, and the crystals of Rochelle salt, Epsom salt, and alum that they grew from solution all have their own kinds of regularities. Perhaps others will find that the cleavage in the mica cannot be repeated in the obsidian (I would not let them try it on my quartz crystals). It may be that one among them may contemplate, as did the French mineralogist Père Haüy in 1781, that the samples before them could be constructed from minute blocks. A sharp observer might note the constancy of interfacial angles in any one crystal. From which discoveries should they be taught? How can this be done?

THE GIANT PROJECT

Sometimes an experiment is largely mental. One day I found a group of eleven-year-old children in rapt conversation. The subject of their intense attention turned out to be one of the monster movies that television programmers discover when all else fails them. Having once been absorbed in the subject of giants myself as a result of a lovely essay by J.B.S. Haldane called "On Being the Right Size," I entered their conversation by asking the question, "Do you really believe that there could be giants of the size in the picture?"

"Why not?" they asked.

"Because there are complications with increasing size," I replied.

"What about elephants?" one lad challenged.

"And dinosaurs?" another added in support.

"The dinosaurs are gone, though we don't really know why. I suppose that their size didn't prove a sufficient advantage. It might have been a disadvantage. Elephants have obviously succeeded, if size is a mark of success. What is interesting is that the ancestors of the elephants we know today weren't nearly so big. Our present elephants are only about a million years old. We do not know whether increasing in size will eventually prove to have been helpful. If you count the age of animals on earth as a mark of success, then the little limpet comes out way ahead. He and his kind have been around for more than three hundred million years, and he doesn't hold the record. A worm, a jellyfish, or a sea anemone is even older."

"But we were talking about size, not age," a young lady reminded me.

"Why can't there be giants like in the movies?"

"Yeah, how come?"

"Instead of a long-winded answer, how about doing something to find out?"

"Do what?"

"Suppose you start by considering a man ten times the size of a fully grown six-foot-tall man—and remember, that means ten times in more than one direction. You can divide the job into small parts and give everyone who wants one a part to do. One team can worry about what problems the increase in size poses for support. What other problems occur to you?"

"How about food? How much will he have to eat?"

"All right. Suppose that your team concerns itself, at least at first, with food as a source of heat. You may find it useful to compare the food intake of our mice with your own. What else?"

"Air. What about breathing?"

"Good. We have solved the problem with the help of a hundred square yards of lung. The insects manage without lungs altogether. It would be interesting, if you get to giant insects, to know if their particular way of getting the oxygen they need sets a limit on their size. Any other ideas?"

"Suppose he falls. You can drop an insect and nothing happens. Even our mice don't seem to get hurt when they fall, but I wouldn't try it with my dog."

"How come? I thought that the Galileo thing showed that . . ."

"Same time, not the same force. Got it?"

"If you recruit anyone else to your project, or if you want to go on with something else when you have finished, think about the sense organs—the eye, for example. Is good vision related to the size of the eye? The largest animals on earth have eyes that are not much larger than ours. Is there a maximum useful size? How about athletic competition? I thought surely someone would want to know whether your giant could tear through the Olympics or run rings around Lew Alcindor. Find out what you can."

Both mathematics and model building were useful in the course of the "giant project." The central relationship was between volume and surface—weight and skin surface, air and intake surface, food and gut surface, weight and bone cross section. If weight support is a function of the bone cross section, this might be tested with dowels. Most of the quantitative findings were expressed in graphs.

Many experiments resulted from an application of assertions by authorities. One group tried to estimate the rate of gas diffusion by tracking odors. The results were not precise, but they suggested that gas diffusion is a slow process.

Finding areas was largely a test of patience and care in the use of tape measure or ruler, but finding volumes inspired some delightful innovations in technique. How does one calculate water displacement for a mouse that floats? One way is to weight the mouse with a rock so that both sink and then subtract the volume of the rock from the total volume displaced. But that way is cruel, so a snug diving suit for the mouse was devised from plastic wrap with an attached snorkel-type breathing apparatus. The suit was weighted with a rock, and the volumes of the rock and the submerged breathing tube subtracted from the total.

Finding the cross section of bones was laboratory work. Finding the diameters of rods and cones in the retina and the length of average light waves was library work.

A statement by Edith Biggs and James Maclean in *Freedom to Learn* (1969) is very relevant to the experience of these children in the "giant project." "Children will enjoy mathematics because it will have relevance to them in the lives they are living; they will appreciate the power and beauty of mathematics because they can use it to communicate their ideas." Their science experiments revealed to the children the relevance, power, and beauty of mathematics. They had learned when they were little to count and to measure, frequently using arbitrary units. They used the results of their counting and measuring in their experiments, and they learned to imagine helpful experimental designs in accordance with what they might count and measure.

Facts and skills take new life when they become instrumental in contributing to intellectual power. It is natural that this should be so. There is nothing in information about the distance to the center of the earth to make the spirit leap of any but a budding Jules Verne. The Pythagorean theorem and extraction of square roots do not inspire most of us when they remain abstractions. But if this information can be used to measure the distance to the horizon whenever we know our altitude, then facts become a means to an end and a new spirit has been breathed into them. Morris Kline

Head and Hand

writes in *Mathematics and the Physical World* (1963) that "mathematics is valuable because of its contributions to the understanding and mastery of the physical world." The children's experience in science confirms that opinion.

ACTIVITY IS CENTRAL

Activity which lets children make things happen contributes to their conviction that learning is possible, worthwhile, and pleasurable. Experimenting builds a sense of control over events and combats cynicism. In those science education programs in America which are suited to the needs of growing children, the children begin to engage in experiments from their earliest days at school, and engaging in them continues to be a central school activity for as long as they learn. They may begin by surveying class foot-size, height, weight, family-size, and the rates of plant growth, and they graph their results. As they grow older they can formulate their own law of the see-saw and investigate Noah's housing problem, limiting their calculations to mammals and their food supply. When they are almost ready to leave, they find Riemann's geometry sensible on the globe, and they learn to draw the hour angles on their sundials more accurately by using Tan (shadow angle) = Sin (altitude) x Tan (hour angle), looking up the values for tangent and sine in the tables.

Children may do these things or a thousand other ones. To experiment is to be uncertain. If the children's learning and the teacher's teaching are to be effective, both must be honest, and if they are to be honest, we must allow for uncertainty in the children's efforts to understand their world, and in our efforts to help them do it successfully.

6. Head and Heart

> "Beauty is truth, truth beauty,"—that is all
> Ye know on earth, and all ye need to know.
>
> John Keats, "Ode on a Grecian Urn"

The esthetic experience in science education is badly neglected. The neglect is understandable, but unfortunate. It is the logical consequence of the conception of artistic experience as peripheral to life and of scientific activity as the exercise of skills. If one subscribes to this view, then one believes it is the obligation of adults charged with the responsibility of educating children in science to help them master the skills—observation, measurement, analysis, etc. Presumably the mastery of these skills will introduce children to the world of science. Unfortunately, when the instruction based on this philosophy is effective, children learn the skills but remain innocent of any understanding of science. The fact is that although science activity primarily satisfies the need to know, it also satisfies the need to feel—to feel order, simplicity, elegance, beauty. Children, like adults, have both these needs.

Early evidence of man's struggle to satisfy these needs can still be seen today in caves in southern France and northwestern Spain and on the walls of Zaraut-Sai Gorge in Uzbekistan in the form of paintings of the bison, wild cattle, the horse, the woolly rhinoceros, the mammoth, the wild boar, and the bear. About 40,000 years ago, during a period in the last glaciation, men learned to make fine-blade tools from flint and obsidian, and knives, pins, needles, and fishhooks from bone. At the same time, they learned

Head and Heart

to visualize in their mind's eye a two-dimensional representation of the three-dimensional animals they saw and hunted. They recorded their vision in paint and line so satisfying that their paintings are still thought beautiful today. In dangerous and uncomfortable places, these men explored new realities, not for themselves alone, but to enable all the members of their community to know better what kind of world they lived in. Life was too hard and uncertain to permit the luxury of art as a hobby or a distraction. Art was useful in the deepest sense of that word, for it affirmed that men's decisions can affect both gods and men, can change circumstances to accord with men's needs, hopes, and dreams. Art has never lost this usefulness.

In that dim world of long ago, men exercised their dawning intelligence to capture for their own use the powers of their dreaded enemy—fire. We shall never know at what precise time or under what circumstances men first imagined that they might harness this power by rubbing one piece of wood against another. It is difficult to believe that an achievement which would extend man's range to all corners of the globe was the result of chance alone. Whether by sawing against the grain of wood, by ploughing along it, or by drilling into it, men in different places discovered that when wood rubs against wood, enough heat can develop to set the wood dust smoldering, and it requires only a man's breath to blow it into a living flame.

The order of the development of art and science is here reversed. Surely the skills which produced the sharpened piece of wood or quill that served as brush, the moss or fur with which to fill the large spaces, and the iron and manganese oxide pigments with the vessels in which to store them must have evolved before the paintings could be attempted. Even more critical in establishing the proper sequence of events is the artist's need to eat. Someone had to supply him with food while he painted, sometimes two miles below the surface. It became possible to feed him without requiring his company in the hunt or in food gathering when techniques were sufficiently advanced. Mine is therefore a distorted perspective. It is so by design. Books about science for children rarely consider art and esthetic experience germane to their purpose, except possibly to the extent of a marginal notation that science is a creative activity. I have distorted the historical perspective, attempting by this crude device to draw the reader's attention to art, as Henry Moore draws our attention (infinitely more successfully) from bony structure to surfaces by distorting his figures.

In those primeval days of cave paintings and fire-making, and for a long time after, science and art were not differentiated. Together with religion they were all contained in magic. The origins of the words "poet" and "artist" suggest that the meanings once associated with them were quite different from our contemporary meanings. "Poet" comes from a Greek word which meant "maker," and "artist" comes from a word which meant "to devise or arrange," which also gives us the word "artisan." Artists were people who made things, arranged them, and put them together. They shaped things, fitted and joined them—clay, stone, space, words, and tones, and thereby gave them form. They sensed their world, and with the gift of imagination, they deliberately set about to modify it in ways not given by the senses. They recognized no hierarchy among objects, feelings, or events. When these efforts, these experiments, were successful, without stated morals or lessons, they enlarged the mind and refined the feelings. They illuminated both external and internal reality. The artist was also the scientist.

THE SEPARATION OF INSEPARABLES

With the invention of agriculture and the advance of technology came the growth of towns and cities. As life grew in complexity, it became more efficient to divide the various tasks. It was not possible for the same men to quarry stone and build statues so some men became quarriers and others became sculptors. The digger of

Head and Heart

stone had only to know how to extract the stone. He was not concerned with its beauty. That was the sculptor's domain. Human activity was divided into matched halves—to act so that one might know and to act so that one might be. To know was to be able to plan future actions; to be was to be able to assess the meaning of existence and decide what must be done.

Actually, life is far too interesting to allow such neat divisions. In practice the either-or separations are not accurate descriptions of the real world. In *Science and Civilization in China* (1956), Joseph Needham describes an ancient Chinese belief in two forces, polar opposites like darkness and light, that governed the world. They called these forces Yin and Yang. They knew that "The Yin and the Yang reflected on each other, covered each other and reacted with each other." In the *Ta Tai Li Chi* (Record of rites of the elder Tai) is written: "Man alone comes naked into the world; [this is because] he has the [balanced] essences of both Yang and Yin." The physiology of man's body confirms that opposite forces are at work. He should be in serious difficulty if anabolism did not balance catabolism to maintain a steady concentration of glucose in the blood. Metabolism consists of the chemical transformations produced by the activity of living tissue. It requires both anabolism and catabolism, a building up and a breaking down. Other pairs such as mass-energy, space-time, particle-wave, love-hate, appear separable into single autonomous concepts, as though one excluded the other. When they are understood, it becomes clear that each implies the other, cannot exist except in relation to the other, and that their relationship to one another is not a placid, stable one, but a dynamic one in which equilibrium is tentative. The realization of the interconnectedness between apparently separate members of such a pair has marked some of the highest achievements in science. There was a law about the conservation of energy. There was also a law about the conservation of mass. Einstein's Special Theory of Relativity brought more understanding of both mass and energy by establishing their relationship. It is not possible to be nearly as precise, nor is it possible to conceive of the regularities implied by laws, but in principle, the division between artist and scientist, between thought and feeling, between thought and morality, are other instances of the separation of inseparables.

Linus Pauling is not an artist and Pablo Picasso is not a scientist —considered only superficially this trivial statement does not merit argument. But to someone who is not content with labels, to someone who wants to know *how* Linus Pauling developed his theory of bonding, *how* Pablo Picasso happened to paint *Guernica*, it becomes apparent that Pauling and Picasso have a great deal in common, and unless their similarities are explored, as well as their differences, it is very difficult to understand either man.

Picasso—Fifty Years of His Art (1946) by Alfred H. Barr Jr. includes an interview Picasso gave in 1923 in which the great artist stated his view of the function of art: "We know that Art is not truth. Art is a lie that makes us realize truth, at least the truth that is given us to understand. The artist must know the manner whereby to convince others of the truthfulness of his lies." In Picasso's painting *The Race,* the movement of massive figures appears to shake the earth, but the viewer's attention is not drawn to the structure and articulation beneath the flesh. The preoccupation with bones and sinews that fascinated the artists of the Renaissance is not shared by the artist in the day of the X ray. Picasso's "lies" tell of a new age, one devoted to the summing of the many particulars into large patterns. In an interview shortly before the Spanish Civil War, Picasso said: "One cannot go against nature. It is stronger than the strongest man. It is pretty much to our interest to be on good terms with it! We may allow ourselves certain liberties, but only in details." In *Guernica,* which Maurice Raynal has called the apocalypse of our time, there are "small liberties," but the horror of war is true, and the outstretched arm holding a lamp for hope and enlightenment is also true. The truthfulness is emphasized precisely because Picasso "lied" to convince us of his vision of the agonizing, brutal, inhuman character of the war against his people and the hope for an enlightened mankind.

Head and Heart

It is the artistic vision which senses the true character of the time in which the artist lives, whether in the spatial contours of the women's figures or the horror of war. Artists have their way of seeing truth.

James D. Watson includes an account of one of Pauling's lectures in *The Double Helix:* "Pauling's talk was made with his usual dramatic flair. The words came out as if he had been in show business all his life. A curtain kept his model hidden until near the end of his lecture, when he proudly unveiled his latest creation. Then, with his eyes twinkling, Linus explained the specific characteristics that made his model—the a-helix—uniquely beautiful." Watson is drawing upon the impressions of Jean Weigle, a Swiss scientist. "Several of his [Jean Weigle's] younger friends ... trained in structural chemistry, thought the a-helix looked very pretty. The best guess of Jean's acquaintances, therefore, was that Linus was right." Picasso is concerned with truth in paintings, and scientists are impressed by a beautiful scientific model. It is the scientific vision of the graininess that forms the basis of aggregates with various degrees of stability that provides our understanding of proteins and stars. Scientists have their way of seeing truth.

It is a mistake to think of the two visions, the artistic and the scientific, as totally isolated one from the other. They are both rooted in the imagination. The truths to which they both aspire are generated in action. The "it" that Shelley speaks of in "it renders the mind susceptible to hitherto unapprehended combinations of thought and feeling" is appropriately applied to scientific activity. Both artists and scientists experiment, and artistic discoveries are made, Milton tells us, "not without dust and heat." (The quotations from Shelley and Milton appear in Arnold Kettle's *Man and the Arts in a Changing World* [1968].) Man has two hands, not one. Some people are right-handed, but this does not mean that their left hand is unused, nor does the left-handed man question the usefulness of his right hand. Analogously, all men think and feel. To reduce them to thinkers only is to make abstractions of men, walking equations. To reduce them to feelers only is to make men another kind of abstraction, physiological systems.

Artists and scientists both express a world which at its material base is modular. The graininess of our material world, which Philip Morrison describes so beautifully in "The Modularity of Knowing" (in *Module, Proportion, Symmetry, Rhythm* [1962], edited by Gyorgy Kepes), is combined in increasingly more complex structures. The bits that can be counted and measured combine into a world that is filled with ambiguities, a world in which man is left to make choices. The artist imagines segments of this totality with its ambiguities, and presents them for us to respond—to sense the light, sound, shape, the form into which he has cast his experience, and to make our individual choices. The scientist is drawn by various portions of those grains to find their interconnections, to find explanations that will account for their behavior on the largest possible scale with as little ambiguity as possible, in a single judgment, if he can.

Poets and playwrights show us a world which is absurd. Social scientists attempt to find the mechanism that makes it so. If they are successful, both contribute to the effort to make it less so. In "The Lesson" (1967), Miroslav Holub, who is both a scientist and a poet, observes and makes an accusation:

> A tree enters and says with a bow:
> I am a tree
> A black tear falls from the sky and says:
> I am a bird.
>
> Down a spider's web
>   something like love
>   comes near
>   and says:
> I am silence.
>
> But by the blackboard sprawls
>   a national democratic

>   horse in his waist coat
>   and repeats,
>   pricking his ear on every side,
>   repeats and repeats
>   I am the engine of history
>   and
>   we all
>   love
>   progress
>   and courage
>   and
>   the fighter's wrath.
>
>   Under the classroom door
>   trickles
>   a thin stream of blood.
>
>   For here begins
>   the massacre
>   of the innocents.

That is the poet's judgment.

Race and Democratic Society (1945) by Franz Boas contains the judgment of a great anthropologist: "I believe that the purely emotional basis on which, the world over, patriotic feelings are instilled into the minds of children is one of the most serious faults in our educational system, particularly when we compare these methods with the lukewarm attention that is given to the common interests of humanity."

Two people, both black, one a poet, the other a social psychologist, respond to the experience of oppression. First, poet Countee Cullen in "Incident" (1925):

>   Once riding in old Baltimore
>   Heart-filled with glee

> I saw a Baltimorean
> Keep looking straight at me.
>
> Now I was eight and very small
> And he was no whit bigger,
> And so I smiled, but he poked out
> His tongue, and called me, "Nigger."
>
> I saw the whole of Baltimore
> From May until December
> Of all the things that happened there
> That's all that I remember.

Now, Kenneth B. Clark in *Prejudice and Your Child* (1963): "The humiliation of any single child in a very real sense robs every other child of some of his humanity.... The maintenance of rigid patterns of racial segregation within the community makes difficult, if not impossible, significant changes in racial attitudes of children. More and more attention must be paid to such urgent community problems as the abolition of residential segregation, equality of opportunity for employment, and the breaking down of all arbitrary racial restrictions that interfere with the general and political freedom of individuals."

The poet and the astronomer live under the same sky. Each ponders its mysteries in his own way, Mark Van Doren in "The God of Galaxies" (1953):

> Now streams of worlds, now powdery great whirlwinds
> Of Universes far enough away
> To seem but fog-wisps in a bank of night
> So measureless the mind can sicken, trying—
> Now seas of darkness, shoreless, on and on
> Encircled by themselves, yet washing farther
> Than the last triple sun, revolving, shows.

Head and Heart

and Carl Sagan in *Intelligent Life in the Universe* (1966): "Is the universe finite or infinite? Is it eternal, or did it have a finite beginning in time? If it was created at an instant in time, how was this accomplished? If the universe is infinitely old, is there a sense in which it has a purpose? Are the physical laws fixed, or do they alter with time? What determines the physical laws? Does the universe have the same appearance at all places and times? What is its geometry? Why do the galaxies seem to be flying apart, one from another? Is there an overall irreversible conversion of hydrogen to the heavy elements? The cosmologists grapple with many of these problems; eventually they may be solved."

Each pair, Miroslav Holub and Franz Boas, Countee Cullen and Kenneth Clark, Mark Van Doren and Carl Sagan, shares thoughts and feelings. The poets have mastered their knowledge, which they express as feeling. The scientists have mastered their feeling, which they express as knowledge.

CHILDREN FEEL AND KNOW

It may well be that our efforts to educate children in science have lacked spirit because we do not assign significance to children's artistic experience. Poetry is frequently something to "translate into your own words" or to break down into the various devices employed by the poet. The novel, more often than not, is "the main characters," "an interesting episode," and one more title to be added to the cumulative reading record. The visual arts and music are "minor" subjects. The drama is to be read and sometimes produced for assembly programs, and the dance is nowhere to be seen. Granted that this description may be unjust to some programs, but where is the effort in the arts to match that of the National Science Foundation? The absence of a comparable art foundation is a reflection of the low esteem in which art and art education are held. Art is a bauble, a distraction from the serious affairs of the world. So it would seem. Art has lost its usefulness. This is how it is perceived outside and inside the school.

If children are to be educated in science, educators and parents must recognize that science contains an artistic component, one that draws on the imagination and the self, a component that cannot be reduced to packaged procedures. Lord Kelvin's old stricture about what is and what is not science is not true about contemporary science, nor is it helpful in educating children. Kelvin's dictum, "If you can measure that of which you speak, and can express it by a number, you know something of your subject. If you cannot measure it, your knowledge is meagre and unsatisfactory," a model of science drawn from nineteenth-century physics. Certainly, as knowledge increases in precision, the underlying mechanisms of the natural environment yield to mathematical expression. But mathematical expression of itself does not illuminate the world any more than the use of logic of itself tells something true or important. Lord Kelvin's estimate of the earth's age was not as good as the estimates by Lyell and other geologists who had no means at arriving at a discrete number. Whose knowledge was meager? In what numbers did Darwin express his theory of evolution? What numbers do the classifiers and ecologists employ? Is anthropology meager knowledge? Field anthropologists find that one good informant may reveal the life style of a people better than any sampling process. Economists employ mathematics, and even computers, as do sociologists and urban geographers. Have these fields produced more substantial knowledge than anthropology in a comparable period of time?

When children are able to, they should certainly be encouraged to count and to measure, but the ability to quantify ought not to be the sole criteria of the worth of their efforts. What is to be thought of the little boy's discovery of the isotropic projection of a sphere? And what of the discovery that acoustic tiles will not accumulate sound? Are these discoveries of little consequence because the children know no way to assign them numbers? To say "yes" is to dehumanize science, to make science fit only for the bone-dry stereotypes too many Americans imagine scientists to be.

Head and Heart

The central problem is not to teach children to assign the proper number to an experience. The real challenge is to cultivate an air, a climate, in which children's imagination is sensitized to conceive of what the senses do not convey, to have wild, exciting ideas, the kind that cause a halt in the ongoing transactions because flights of the imagination appear unrelated, at least for a little while, to all that has gone before. Essential ingredients for such an environment are mutual respect and patience. It is an atmosphere where the child who announces that she has built a pendulum with magnets, a pendulum that will swing forever, is not greeted with shouts and laughter, but with serious and admiring questions. It is where a child is not compelled to witness a dissection, or where a private request for assistance does not become a public announcement. It is an environment which provides time to tinker and to fail, time to chat and to reflect, to stare vacantly while the elusive threads tangled deep below disengage themselves and find their way to consciousness.

It is a happy place not governed by a scarcity economy of good feelings that are the rewards of pleased adults, but one filled with activity, where feelings are associated with the task and the acceptance of one's fellows. It is a place where joys and sorrows, as well as knowledge, are shared. It is a community.

There is no recipe for building such a community. Hughes Mearns, Sylvia Ashton-Warner, Jonathan Kozol, Gerard Rosenfeld, Herbert Kohl, James Herndon, Sybil Marshall, A.S. Neill, A.S. Makarenko, and thousands of teachers all over the world evolved their own unique ways to build one. What characterizes their efforts is a profound respect for children, not mawkish sentimentality, but deep-seated belief in the potentials of human development through learning.

Every successful teacher fashions the circumstances, often against opposition and indifference, that wed art and science and knowing and feeling. A true teacher enables children who are spurred by the human desire to know and feel to reach out and, in the process, to cultivate their humanity. No sequence of courses accounts for such a teacher's devotion, insight, and humanity. Somehow, in the course of varied lives, each awakened passions in behalf of life, and nurtured them with the indivisible trinity of the good, the true, and the beautiful.

7. Error and Failure

> Wise men in long white togas come forward during the
> Festivities, rendering account of their labours,
> And King Belos listens.
>
> O, mighty King, says the first, I've made a pair of wings
> for your throne. You shall rule from the air.—
> Then applause and cheering follow, the man is
> richly rewarded.
>
> O, mighty King, says the second, I've made a self-acting
> dragon which will automatically defeat your foes.—
> Then applause and cheering follow, the man is
> richly rewarded.
>
> O, mighty King, says the third, I've made a destroyer
> of bad dreams. Now nothing shall disturb your royal sleep.—
> Then applause and cheering follow, the man is
> richly rewarded.
>
> But the fourth man only says: Constant failure has dogged
> my steps this year. Nothing went right. I bungled everything
> I touched.—Horrified silence follows and
> the wise King Belos is silent too.
>
> It was ascertained later that the fourth man was
> Archimedes.
>
> Miroslav Holub, "Inventions"

The textbooks have distilled most of the failures from the history of science. There is little else that can be done; including them

Teaching Science to Children

would fill entire shelves. But adults working with children in science must know that for every success there are countless errors and failures. The greater the achievement, the greater struggle required to make it possible. This is one of the lessons one learns from the history of science. If children are to learn science, that lesson must become operative in their science activities.

To say that error and failure are normal in scientific work is not to denigrate science. Nor is it to add to the metaphysical speculations about the limitations of science. It is merely to say that error and failure are normal to men at any moment in time. Anyone who has tried to *do* science knows that it is not possible to avoid mistakes. People who have a secondhand acquaintance with science, who know it only from textbooks or by reputation, think of scientists as infallible practitioners of arcane activities. In this view, science is rather like a secular faith whose adherents are all popes. If this were true, the prospect of studying science would be frightening and forbidding, as indeed it is in the minds of many. Happily, it is not true at all, nor should it be surprising that scientists, even the greatest scientists, make mistakes. The only way to avoid them is to avoid new thoughts—but then there could be no science.

INSUFFICIENT INFORMATION

Some mistakes are inevitable because not enough is known; the information required to avoid them is not yet available. Aristotle believed that in order to keep a body moving, a force had to be continually applied. If the action of a force was not evident, then air particles maintained the action. According to this view, if someone were to set a ball rolling, air particles would rush into the space behind the ball after it left his hands and keep it rolling. It was not until the fourteenth century that a French philosopher, Jean Buridan, reasoned differently. He felt that the first push was enough —God set the heavenly bodies in motion and that was sufficient

Error and Failure

to keep them moving. Buridan anticipated Galileo's work and Newton's First Law, and Aristotle was wrong.

Sir William Herschel, the most important astronomer in the late eighteenth and early nineteenth centuries, showed that the solar system was a tiny spot in a huge star system, but he thought that the solar system was in the center of the galaxy. He also held strange notions about the sun. He believed that the sun's light and heat were only on the outside. Below this shell of light the sun might be cold and perhaps inhabited. The sun spots were the holes through which the surface below could be seen. These notions were all erroneous.

INTELLECTUAL INERTIA

Though the lack of sufficient information is probably the greatest single cause of error and failure, it is not the only cause. Scientists, like other people, find it difficult to part with cherished beliefs. The unwillingness to accept a new idea is often a source of error.

Jean Louis Rodolphe Agassiz, the nineteenth-century Swiss-American naturalist who made notable contributions to the understanding of glaciers and fossils, and his contemporary Claude Bernard, the great French physiologist, were antagonistic to Darwin's theory of evolution. For a time, Charles Lyell, who contributed uniformitarianism to geological theory, also found it too threatening to orthodox belief to accept the possibility that man, too, might have evolved from lower animals. Darwin replied to Lyell's objections and convinced him—no mean achievement, for Lyell was recognized as England's greatest geologist, botanist, and zoologist. Sir Gavin de Beer includes in *Charles Darwin, a Scientific Biography* (1965) a letter that Darwin wrote to Lyell on the day before the publication of *The Origin of Species:* "To have maintained in the position of the master, one side of a question for thirty years, and then deliberately give it up, is a fact to which I much doubt whether the records of science offer a parallel." Indeed, it is very rare.

In 1811 Amadeo Avogadro, an Italian physicist, proposed the hypothesis which bears his name. According to this postulate, equal volumes of all gases, under the same conditions of pressure and temperature, contain the same number of molecules. Carbon dioxide has a molecular weight of forty-four. Forty-four grams of carbon dioxide contain Avogadro's number of molecules, 6.0238×10^{23} molecules (602,600,000,000,000,000,000,000). This number is known to be correct within less than 0.01 per cent, a greater accuracy than the census figures for any city with a population greater than 10,000. It was a brilliant piece of work, but John Dalton, an Englishman, and Jöns Jacob Berzelius, a Swede, giants of chemistry in the first half of the nineteenth century, and other famous chemists of the day, either rejected or ignored Avogadro's hypothesis. Almost fifty years later, Stanislao Cannizzaro, an Italian chemist, struggled to have it accepted and succeeded. It is interesting that Cannizzaro, who battled for a worthy but neglected idea, also joined Garibaldi's small army in its attack on Naples for the unification of Italy.

Gregor Johann Mendel, the Austrian botanist who laid the foundations for the science of genetics, did not fare any better than Avogadro. In 1865, when he delivered his paper on his genetics research, there were botanists, a chemist, an astronomer, and a geologist in the audience. Nobody asked him any questions. They listened politely, and after a proper interval, they left. Perhaps these were not very talented people. However, Karl W. von Nägeli of Switzerland was the leading botanist of his day. When Mendel sent him his paper, Nägeli ignored it. Like Avogadro, Mendel had to wait some fifty years to be rediscovered.

There are many more examples of errors caused by the refusal of scientists to change their minds. Priestley and Cavendish refused to give up their explanation of burning although the evidence for a better explanation was available to them. Rudolph Virchow, the German scientist of the nineteenth century who

Error and Failure

founded cellular pathology, refused to accept Pasteur's germ theory. Rather than accept it, he left the field of biology altogether.

Nor are the mathematicians immune to this kind of error. In 1811 Jean Baptiste Joseph Fourier, a French mathematician, submitted a paper to the Paris Academy. The readers and critics were Pierre Simon Laplace, Joseph Louis Lagrange, and Adrian Marie Legendre, three giants of mathematics. They criticized the paper so severely that it was not published at that time. The paper on the propagation of heat has become a classic.

SNOBBERY, SENILITY, AND CERTAINTY

James Prescott Joule, an English physicist of the nineteenth century, was the first to determine the mechanical equivalent of heat. He found that a certain amount of work always produces a certain amount of heat. But Joule was a brewer, not a professor. This seems to have been held against him, because the learned journals rejected his original statement and he was never given a professional post. Only Lord Kelvin's interest finally obtained him a hearing.

In 1802 William Hyde Wollaston, an English chemist and physicist, discovered that the spectrum of sunlight was crossed by dark lines. In 1814 Joseph von Fraunhofer, a German physicist and optician, rediscovered these lines, and by increasing the dispersion with the use of more than one prism, he observed many more lines, and mapped them carefully. Eventually his discovery led to spectrum analysis of chemical elements and to spectroscopic astronomy. The scientists of his day denied Fraunhofer the status of colleague and considered him a technician. He could attend their meetings, but he could not speak.

In 1813 the London College of Physicians denied membership to Edward Jenner, the discoverer of a vaccination against smallpox, the most dreaded disease of his time, because Jenner would not let himself be examined on Hippocrates and Galen. He thought that his discovery was sufficient qualification.

These errors in judgment appear to be the result of ordinary snobbishness. Sometimes errors are prompted by the indisposition of old age. Marie Eugène François Thomas Dubois, a Dutch paleontologist, discovered a skull cap, a femur, and two teeth, the remains of "Java man," *Pithecanthropus erectus*. He published the details of his find in 1894. The evidence for the evolution of man created great controversy. Subsequent finds in China and Africa gave strong support to Dubois' find as a "missing link." After all this had been well established, Dubois, then an old man, insisted that the remains were of a fossil ape.

Expressions, even by the greatest scientists, that set limits on scientific discovery, eventually prove to be erroneous. Huygens thought that six planets and six satellites make for a harmonious number. He predicted that more would not be found. Rutherford was doubtful whether man could control the energy of a atomic nucleus. Social control is still in doubt, but Rutherford meant physical control, and that has been achieved.

MAKING MISTAKES

Children will make all these kinds of errors except those that result from the deterioration of aging. A very common children's error is caused by crude experimental design. They are in good company here, too. Richard Willstätter, a German scientist, received the Nobel Prize in 1915 for his work in chemistry on plant pigments. In the 1920s his research was on enzymes. He announced that enzymes were not proteins. This was believed for ten years. Then James Batcheller Sumner and John Howard Northrop, both American biochemists, showed that enzymes are proteins. Willstätter's tests were not refined enough.

Children should expect to make errors and to fail, and adults who undertake to teach them science should not prevent their

Error and Failure

doing so. Preventing error and failure denies children the opportunity to learn from their mistakes.

What they need to learn is not how to avoid mistakes, but how to handle them. A good way to teach this lesson is to organize experiences in which it is entirely dignified and not at all humiliating to say "I don't know." If young children are shown shadows, but not the objects creating them, and are asked to figure out what the objects may be, their first inclination is to guess. They soon discover that shadows are tricky. What looks like a block of wood may turn out to be a book. A thin line may turn out to be a phonograph record, and a circle, a cylinder. After having been fooled once or twice, some children begin to hesitate before naming the object that is held in the light. They say that they are unsure and ask that the object be turned. At this point a new rule is added to the game. The children may request that the object be moved any number of times provided that they preface each request for change with the statement "I don't know." If they choose to name the object, and they are mistaken, they are out of that round. It is a joy to see children stand erect, eyes twinkling, mouths smiling, saying, "I don't know." They attach no shame to the statement. Their smiles suggest, "I don't know yet, but I'm not going to be fooled so easily. There are things I can do that will give me a better clue. I'm pretty smart."

The objective here is not to introduce projective geometry, although the children are bound to learn something about it, but to help children devise strategies in attacking puzzling situations. The aim of this experience is to help children learn that in the absence of adequate information, it is not possible to know. In such a case, it is best to circumscribe ignorance by squarely recognizing that one does not know and developing strategies to find out. It was in the course of this game that a little boy announced, "I made a thing out of clay. No matter how you turn it in the light, it always makes the same shadow."

Any game that requires children to deal with ambiguity serves the purposes of the shadow game. An array of appropriate materi-

als may be presented to children, and each child in turn required to make a choice among them. For example, if there are cubes, spheres, cylinders, pyramids, and cones of various sizes and colors arranged either in a line, a square, or other configuration, a child may mentally choose an individual piece and identify the choice by telling one other child, "I'm thinking of the littlest red cube." The child whose turn it is to be the questioner names each shape available and asks whether it is that one until the "thinker" tells him he has picked the right shape:
"Is it a sphere?"
"No."
"Is it a cone?"
"No."
"Is it a cube?"
"Yes."
Now, the problem is to ask the kind of question that will limit the possibilities: the strategy is the same as that used in playing twenty questions. The point of the game is to see how few questions are required to identify the particular piece. Of course, if there is more than one little red cube, the location of the particular red cube will have to be given in reference to its neighbors. It can easily be seen how the game will progress as the children become more familiar with the shapes, and how it may be adapted to other domains. In these games children have no option but to do their own thinking, to deal with ambiguity, to make mistakes, and to learn to correct them.

Doing one's own thinking often leads to error but what other kind of real thinking is there? What is gained by the repetition of other people's words if one cannot repeat the thought processes that led those people to the words? When a child searches his own thoughts, he will often come to incorrect conclusions, but in the process he will have grown in independent intelligence, the sine qua non for discovery, creative work, criticism, and dissent. If he is going to risk being wrong, he has to find that it is not a humiliating, fear-ridden experience. He needs circumstances that encour-

Error and Failure

age him to try again, to do something differently, to seek more information, to turn himself around, but to continue to "do his thing."

Children should not be sheltered from making mistakes except those that clearly threaten their safety or health. Even in matters of children's well-being it is sometimes better to take a calculated risk than to condemn them to passive sterility. It is the problem every parent faces when a young child wants to climb a tree or a jungle gym, or when he wants to cross the street by himself. If he never climbs, he will not fall; if he is not allowed to cross the street alone, he is far less likely to meet with an accident. Similarly, if he never learns about the wonders of electricity, he is less likely to try dangerous experiments. But what kind of person will result from such suffocating protection?

Children need to learn to do things with minimum risk. Should they make a mistake, let it be one they can correct. They must learn that the errors to avoid are those that eliminate opportunities to try again. It is a lesson they will continually struggle to master in the context of their contemporary world, as we who instruct them struggle with it in our deliberations about weaponry and security, expenditures and world famine, race and social harmony.

Thus it is best to let the nine-year-old girl make the bottle barometer she saw illustrated in a book. Let her find out for herself that it is useless except under conditions of constant temperature. The point is, having found out, what will she do? Will she choose a vacuum-sealed can and labor over the difficulties of calibrating it?

Let the eleven-year-old boys try to generate hydrogen, fill soap bubbles with it, and light the rising bubbles—just be certain that the vessel in which they are producing the hydrogen is small. When they bring the flame too near the generating apparatus, there will be a loud bang, but it will not hurt anyone. And it will teach them more about the need for careful laboratory procedures than any amount of warning.

If the ten-year-old girl thinks that the string she is laying along the perimeter of a simple irregular closed curve will help her to find its area, do not dissuade her. It is an ingenious idea, but it will not work.

Apparently none of the eight-year-old children engaged in making a model of the earth has ever made a beanie or noticed the lunes in an umbrella. The paper with which they are trying to cover the lath ribs does not fit without wrinkles and folds. Let them struggle.

The working model of the Foucault pendulum is not working satisfactorily. It did not swing for a sufficiently long time, nor did it leave noticeable scratches in the sand on the floor. Perhaps the children are not interested enough to bother to overcome their problems. Telling them the solution will not be of much value.

There is an old story about an astronomer's child. One evening when Father had already gone to his work and his son and wife were having dinner, the boy asked his mother to help him with his homework after dinner.

"What do you have to do?" asked Mother.

"I have to give a report on the moon," the boy replied.

"How silly to ask me," Mother said, "when your father is an astronomer."

"I don't want to know that much," the boy retorted.

The eight-year-old children will learn as much about the Foucault pendulum as they can at this particular time. It is the adult's responsibility to point out the pendulum at the United Nations and to provide them with materials, time, reference books, and a climate in which they can try, make mistakes, recover, push on, and quit when they have done what they can do.

USING DISAGREEMENTS

Sometimes children disagree about what ought to be done. The ensuing controversy is frequently very instructive. A group of kindergarten children were arranging a collection of boxes and bags

Error and Failure

according to weight. They agreed on the relative weights of all but one. Everyone except one little girl thought that a small box (filled with weights) was the heaviest of the lot. She insisted that the bag (filled with sand) was the heaviest. How to resolve the impasse? "Ask the people," a boy suggested with a wave of his arm. The people were asked and, of course, said what they had said before. The little girl remained a minority of one. It is tempting at such times to give a talk on majority votes and scientific truth. One ought to resist that temptation and let the children grapple with the problem at hand. It became very quiet as the children did their hardest work of the day—thinking. It was an uncommon event for a classroom. The teacher was silent, the children were thinking. Superficial observation revealed nothing of consequence happening; at such moments conscientious teachers worry that time is being wasted. They find silence hard to tolerate and usually cue the children to get them going in the "right" direction. We simply waited. Suddenly, a boy shouted "POUNDS!" so loud everyone jumped. He had just found a use for the scale.

The lone dissenting voice turned out to be correct. It was a shocking experience, more especially because the little girl, who published her own weekly newspaper, was not at all popular in the group.

KNOWING WORDS IS NOT ENOUGH

A variety of means are employed to avoid the risks of error and failure which accompany thought and activity. One of the favorite substitutes is words, labels. Mysteries are "resolved" by giving names to things. Balls bounce because they are made of rubber, and rubber is elastic. Are steel balls elastic, too? Clay is not elastic, but dishes bounce when they are dropped. Do these words—"rubber," "elastic"—really come to grips with the mysteries? Evergreen trees do not lose their leaves at one time but deciduous trees do. What has the tautology contributed? The satellite tends

Teaching Science to Children

to move off at a tangent to its orbit because of centrifugal force. What is centrifugal force?

Mark Van Doren questions such complacent reliance on words in "If They Spoke" (1953):

> What names? They have not heard the sound,
> Nor in their silence thought the thing.
> They are not notified they live;
> Nor ask who set them wandering.

Words and names do imply a sense of certainty. To name something somehow suggests the acquisition of more knowledge than one possessed before. But knowledge of a label does not assure understanding. He who understands can cause something to happen, can impose change, and can often anticipate results. When only the words are known, the results are always surprising. Children who mix together substances from the labeled bottles in their chemistry sets never know what to expect. Such activity, like the operation on the frog, is satisfying dramatic play, but has very little to do with science.

The penalties paid for confidence in words as a substitute for activity are heavy. One of them is the complacent assumption that events will turn out the way they are "supposed" to. A young woman preparing to become a teacher was student teaching, grappling with the last of a series of obstacles between her and the prize, a teaching certificate. Asked to teach a science lesson in the fifth grade, she decided to do something tried and true. She would demonstrate that the atmosphere exerts enough pressure to collapse a gallon can. Having succeeded in collapsing a can at home, she brought the twisted remains to illustrate what could be expected to result from her efforts in class. The old can sat on a table; next to it a fresh one stood on a hot plate, the screwcap off, the water coming to a boil. The children watched, fascinated. The student teacher was uneasy because her supervisor sat in the back of the room but when the condensation above the can gave evi-

Error and Failure

dence of active boiling, she confidently removed the can from the hot plate, screwed the cap on, and waited. Nothing noticeable happened. She poured cold water on the can. Still nothing happened. The student teacher became frantic—"I don't understand why it isn't working. It worked at home. It's *supposed* to collapse." Then, in utter desperation, she attempted to perform an act of legerdemain. She tried to substitute the can she had brought from home for the can which remained uncollapsed.

The children were not fooled, of course; some shouted "fraud." The poor young woman was almost in tears, the children were laughing, and the supervisor silently left the room.

This was a most unusual performance, but it serves to emphasize a common notion about how problems in science are resolved. "Is this the way it's *supposed* to work?" is a question frequently heard in classrooms. What children need to learn, what the student teacher had not learned, is that terms such as "should" apply only to people. A can does what it does. That this one failed to collapse is, indeed, rather odd, but somewhere there is an explanation for its failure to do the expected. Perhaps it had a leak, or maybe it was unusually strong and could withstand the difference in pressure that was built up. The student teacher did not realize that the puzzling behavior of the can provided an opportunity for a serious investigation, a far better experience than the demonstration she had planned. She had relied entirely on someone else's words about what would happen, and when those words proved wrong she simply said, "I don't understand." She did not attempt, through thought or further activity, to gain understanding.

The disastrous lesson was part of the price the young teacher paid for not having learned that any effort to manipulate the forces in the natural environment is, to a lesser or greater degree, open to alternative possibilities, and that it is the struggle to understand these alternatives in activity that makes science so exciting.

The ability to recite a laboratory manual may prove useful on occasion, but it is obviously no substitute for real understanding or activity.

Teaching Science to Children

We have to be especially careful in science learning that language does not serve as a substitute for exploration. When children explore, they sometimes clarify the problem for themselves if they are given an opportunity to talk about their work. But they ought not to be pressed to do so. To introduce too many words too early may be to reduce the problem to words, and the mystery will no longer beckon. It appears everything is already known, that nothing is left to explore.

A nine-year-old girl was intrigued by vegetable colors—just a few drops colored a large jar filled with water. Given 9,000 as an approximation for the number of drops in a pint, she began an investigation to find out how little vegetable color is required to show a noticeable effect on an entire pint of water. The exploration led her to tiny parts, thousandths of a drop. She made many mistakes in the process, mostly in arithmetic, but the experience was a better introduction to the world of the very small than any amount of talk about molecules.

A group of children back from a winter trip to the country tried to build a beaver lodge from mud and sticks, but their model would not hold together. They failed. It was a salutary discovery that beaver lodges are more complex than they appear. Perhaps another examination, more information, is required. The problem is not resolved by talking about beaver constructions.

Children do not fear error and failure until adults make them something to fear, until they become a source of humiliation to children because they are disappointing to teachers and parents. The adult incursions begin early in the life of children. Whatever freedom children may have experienced in divining their world is soon limited. Explorations lose their openness, their unrestricted flights. It is as though the children had been left with only their right hands. The dreamer has been left behind. So soon.

How free can a five-year-old feel when his mother asks whether the kindergarten provides a good start for entry into one of the preferred colleges? Is that what kindergarten is for? Is the only justification for being five years old that one day one will be eigh-

Error and Failure

teen? Such thinking leads to deferred living. According to this view, whatever we do is for something later. Youth's rebellion against the "onward and upward" philosophy becomes understandable. Climbing without savoring the scene is dull work. It is tiresome. People who make social climbing their life's ambition are tiresome people.

Adults fulfill their responsibility to children when they provide them with opportunities to try in circumstances that do not threaten the children's safety. To try inevitably means to fail sometimes and to be mistaken more often. A young child who builds a tower from clay will be defeated in attempting to make it taller than the design will support. He will sometimes take clay from the base and add it to the top. Now it is worse—the structure wilts. He will either quit or learn alternative ways to provide support. The failures are usually greeted with giggles and laughter. Drooping buildings are funny. And in the absence of adult pressure children eventually learn to build their towers tall and straight.

8. What Have They Learned?

Our researchers into Public Opinion are content
That he held the proper opinions for the time of year;
When there was peace, he was for peace, when there was war, he went.
He was married and added five children to the population,
Which our Eugenist says was the right number for a parent of his generation,
And our teachers report that he never interfered with their education.
Was he free? Was he Happy? The question is absurd:
Had anything been wrong, we should certainly have heard.

         W.H. Auden, "The Unknown Citizen"

There is an ancient technique to assist in recording the judgment of a child's progress in school which can be easily adapted to be useful in evaluating a science program like the one I have described. On the left-hand side of a grid like that in the conventional attendance record, one under the other, record the names of the children. Across the top, above each column, place the name of an experience which has been planned. These may include any of the following: pendulum, frog, magnets, field, balloons, electricity, shadows, mirrors. Then leave some blank spaces for unplanned activities which children may introduce. Past the blank spaces the following are listed: questions-problems, errors-failure, esthetic, cooperation, independence, criticism, originality, honesty, persistence, books, math, manual. When a child gives evidence of the ability to pose questions and isolate problems, to handle error and failure, to strive for and appreciate the elegant, to cooperate, to hold onto his ideas in the face of opposition, to offer and accept criticism, to have ideas of his own, to credit others

What Have They Learned

when their ideas have been used, to stay with a task even when it has become monotonous or trying, to use books skillfully, to use appropriate quantitative techniques, and to make constructions, the appropriate box is marked with an X. The boxes under those objectives which seem to have been learned only in part are marked with a single diagonal. If there has been no evidence to merit a judgment, the box is left blank. Finally, if evidence suggests the opposite of what is considered desirable, n is written in the appropriate space. If a child is free with his criticism, but it is consistently destructive, such behavior is marked $n \rightarrow$, the arrow indicating that the negative behavior is in giving, rather than receiving, criticism. If ideas taken from books or other children are claimed as original, this would be noted as n under "honesty."

This record is not intended for report cards or any other kind of public display. It is designed to record the effectiveness of the science program so that it may be modified in the directions that the assessment suggests may be necessary. Whatever the goals and the means employed to evaluate the extent to which children have attained them, the responsibility for the results is the school's. When a program has been designed to educate children in a particular way, and some children give evidence of not having been so educated, it makes no sense at all to say that these children have failed. It is far more reasonable to conclude that the program has failed with these children. It is almost as ludicrous to blame children for the failure of school as to blame patients for the failures of medicine.

FINDING EVIDENCE OF LEARNING

A sensitive teacher who observes children closely in their activities, listens to them, and observes them in their interactions with other children will have little difficulty in finding occasions to provide evidence for making judgments about the behavioral matters. A child who struggles heroically to locate the keel so that the boat will balance has given clear evidence of his ability to persist in a difficult task. It is a little more difficult for a teacher to create contexts in which children reveal whether they have gained insight from the science experiences in which they have been engaged. What have they learned about the pendulum, the frog, or the other things? Some children will reveal what they have learned very clearly during the sharing sessions. Those who keep a great deal within themselves may reveal very little. Still others, those members of a team who have been assigned small, discrete tasks in the group report, may reveal almost nothing. A group assignment may not give a child sufficient opportunity to demonstrate his understanding. On the other hand, it may be that the group, having recognized the limitations of his understanding, assigned him a modest role. In any case, additional evidence is needed to make a judgment. There are ways to acquire it.

When a child has gained genuine understanding, his ideas are no longer restricted to particular associations, but are freed from them to appear in a new, sometimes novel, guise. Good teachers have always known this and have incorporated the knowledge in their practice. Illustrations from adult experience may clarify how understanding, or the absence of it, may be revealed in changed contexts.

A father who "learned" about spinning tops when he was in school but now urges his little girl to ride her bicycle very slowly did not really understand that the faster a body spins, the more readily it maintains its axis of spin. It requires more skill to balance a bicycle when riding slowly. A mother who "took chemistry" in high school but now raises the flame under the *boiling* water to get more heat did not really understand the nature of boiling.

To find out whether children have learned beyond the specific materials provided for them, it is necessary to introduce change and to alter the circumstances in such a way as to draw on the full range of their understanding. The number of ways that this can be done is limited only by knowledge and imagination. Sharing their work with the rest of the school, or with another group, stimulates

What Have They Learned

some children to tap the limits of their understanding. Most children try to contribute. The quality of the contributions varies—what is ambitious for some is pedestrian for others. I recall a discussion to consider which pendulum experiences might be interesting to other students and whether these experiences ought to be shared with the whole school at an assembly, at a grade level meeting, or by means of displays in the main hall.

"Other things that go back and forth are like a pendulum."

It was a tentative statement, not quite, but almost, a question. I did not say anything.

"Well," she continued, "things that make sounds are like a pendulum. I thought that if that's right, then I could make a scratch record of the pendulum swing on smoky glass the same way I did with the tuning fork."

She did, not on glass, but on a smoky can, because it was easier to keep a can moving steadily than a flat piece of glass.

"If the pendulum comes back because of gravity, then our clock pendulum would work differently in different places."

He was wrestling with Huygens' problem, but no one could think of anything to do.

"Part of our record shows a pattern almost good enough to say with numbers, maybe a formula even."

So it did. It showed that within considerable limits, the square of the period is proportional to the length. It went even better when agreement was reached upon the need for a constant in the algebraic expression.

"Arms and legs swing something like a pendulum. Well, a taller person should be able to run faster, or longer anyway, than a short one."

"That's not right. Look at Jerry, he's not the tallest in the class, but he sure is the fastest."

"That's true, but still, when you go for a walk with your father, don't you sort of have to run a little to catch up?"

"Cheetahs are faster than other animals, but they're not the biggest. Besides, I can run almost as fast as my father. But you're

Teaching Science to Children

right too, because I can't keep up with him when we're just walking. Funny."

There was nothing I or anyone else could think of doing to see what difference there may be between the frequency of forced and free oscillation and the length of a pendulum, so the children were left to puzzle as best they could. Planning discussions are one way to reveal the quality of children's understanding.

A presentation before another group, usually a younger one, is also indicative of what children consider interesting or important. The members of the group in charge of a bullfrog collected their information by observation and reading, divided it among the group, and informed younger children that the bullfrog never seemed to have his mouth open except when he ate. The book said that it was because he had to keep his mouth closed in order to breathe. They had tried to keep his mouth open, but the frog struggled very hard, and they stopped because they did not want him to die. Some of the audience wanted to see them open the frog's mouth. The committee consulted briefly and decided they would. They opened his mouth and placed a pencil across his lower jaw. It was a great success.

The demonstrators then put mealworms in front of the frog. The frog hopped away. When he was brought back, he just sat there. The children watched intently. It was quiet. Suddenly, the frog's tongue darted out, and the startled children jumped and giggled. A worm was gone, to the announcement that "He goes for worms that move." The floor of the room was the demonstration table, but every few minutes one of the "frogmen," as they were popularly known, let the frog rest in a basin of water. They explained that the frog's skin must be kept wet because he breathes through his skin, too, and he cannot do it if it's dry. They splashed water in the basin to show that the frog could hear. The frog turned several times when this was done. They also answered a number of questions. Most of this information had been gleaned from books. A teacher can learn something about children's understanding from such demonstrations.

What Have They Learned

Puzzles can also be instructive to the teacher. An object is hidden in a cylindrical cardboard container. The problem is to decide whether there is anything hidden in the container, and if there is, to extract as much information about it as possible without opening the container. Some children try to find a point around which such a container would normally balance when suspended. Others try a more direct attack by sounding it and by rolling it. One girl tries a magnet despite advice that it is pointless because the container attracts in many places. She does find that the magnet is attracted in many places, but it is repelled in one place. Most children now settle for the idea that there are many iron things and a magnet inside. The girl insists that she is not finished yet. She blocks the container on both sides so that it will remain steady, and sprinkles iron filings over the spot that repelled the magnet. She comes to the conclusion that there is a doughnut-shaped magnet inside with an outside diameter of about one and a half inches and an inside diameter of about one-half inch.

Other containers have nails or bundles of paper clips fastened to the inside. Still others have variously shaped magnets, including magnetic tape. What the children say and do contributes to the feedback that a teacher must receive if she is to make judgments about the effectiveness of prior experience.

TESTS PENALIZE THE THINKER

The central purpose of evaluative procedures is to inform the teacher about how adequate or inadequate her provisions for learning have been. The purpose is not to measure the children because the purpose of education is not measurement, it is to teach. Good evaluative procedures suggest to a teacher what children are ready for, and, therefore, what new provisions the teacher ought to make in their learning environment. Fear-inspiring, competition-generating, formal tests do not contribute to effective teaching or learning, and no one has yet devised means to free tests of these corrosive attributes.

The suggestion to omit formal testing from the school program leaves many teachers and administrators very uneasy. Such people have not reflected upon what of educational consequence has been achieved when the children are tested and their grades dutifully recorded. When they do reflect upon it, they are likely to discover that apart from instilling fear and competition, tests may militate against the cultivation of independent intelligence. In fact, this quality of mind may deny a child a good grade.

Professor Alexander Calandra's story is a good example of what I mean. In a physics examination, students were asked to determine the height of a building, given a barometer. One student, knowing precisely what was wanted in the question, refused to comply. He offered three alternative solutions. The first was to drop the barometer from the roof of the building, time its fall, and by $s = \frac{1}{2}gt^2$ calculate its height. This solution was rejected by the instructor. His second solution was to drop the barometer, measure how long it takes to hear the sound made by the impact, and, using 1090 ft/sec + (2 ft/sec per degree Celsius times the temperature) for the velocity of sound in air, calculate the building's height. This, too, was rejected. Finally, he suggested that if he offered to give the barometer to the building superintendent, surely the superintendent would be willing to give him the information.

The student undoubtedly knew the accepted formula, that for small elevations the barometer falls one tenth of an inch for every ninety feet of ascent. One suspects that he objected to the ritual of one "right" answer per question, that it offended his sense of independence and integrity, and this, of course, is the point the story is intended to make.

Granted that this is a most unusual case, it nevertheless captures the negative aspects of testing. (The offensiveness and futility of testing is thoroughly demonstrated and documented in Banesh Hoffmann's *The Tyranny of Testing* [1962].) At best, a test indicates whether learners have amassed the information and computational skills that a teacher has selected for them to learn. When

What Have They Learned

tests become the measure of success, students are encouraged to second-guess the teacher about what material will probably be covered by the test and to learn it, at least until the examination is over. In this system, what do students learn that is of lasting value to them? What do teachers learn to help them teach more effectively?

TEACHING—NOT TESTING

Tests provide numbers, and that knowledge which can be expressed numerically is presumed to be more significant—remember Lord Kelvin's dictum? The school is not the only institution that has been deluded into accepting this misguided view. The government frequently makes the same mistake. This is attested by Arthur R. Ross, who served as Commissioner of Labor Statistics from 1965 to 1968. In an article entitled "The Data Game," which appeared in the February 1969 issue of *The Washington Monthly*, Mr. Ross writes: "The officials are vulnerable because they are searching desperately for ways to clarify and simplify the protean problems of government. Statistics enable them to do this at the cost of heroic oversimplifications: one or two dimensions, which happen to be measurable, serve to symbolize an elusive, many-sided phenomenon." He continues, "The trouble is that the unmeasured, or unmeasurable, aspects of a problem may be vastly more important than those which have been, or can be, measured."

Measurement is one of man's most powerful tools. It consists of assigning number values in accordance with some expressed rule. What rule of any genuine educational significance justifies the numbers placed on the top of a test? What does 35%, 65%, 80%, or 100% mean about a student's science mastery? What do these numbers say to the student? What do they say to the teacher? They are measures of the response by students to questions about material selected by someone else. In accordance with the number of right and wrong answers they give, students are adjudged excellent, good, fair, poor, failures, or by euphemisms for these which fool no one, least of all the children.

These and similar designations based on tests, even the most meticulously designed tests, are inappropriate estimates of a student's science education. They are fairly good estimates of a student's desire to be successful in school, of his willingness to conform, and of his memory. But they have little to do with the development of women and men of integrity, independence, and imagination who will be able to fashion the world in accordance with their insights and their visions. In conditioning children to accept the attitudes which tests foster, educators imply that the admirable people are the acquisitive, the unscrupulous, the aggressive, the opportunistic, and the idiot savants. Measurement and children are thereby both abused.

Measurement can, indeed, be a useful tool. But when it is employed to demean people in tests, it is abused. Children's efforts in the visual arts, music, the dance, and drama, have not yet been subjected to neat numerical scores. It may be that the testers have thus far neglected the art experiences because these are generally considered to be relatively unimportant peripheral or "fringe" activities. It is also undoubtedly very difficult for the testers to find rules, even primitive rules like those they have invented to discern the acquisition of knowledge, that make any sense at all in the arts. Like compassion, love, tenderness, or honesty, beauty does not lend itself readily to ruler and graph. It is emotion- and value-laden. So are insight and imagination, and what manner of science that is not a caricature is learned without them? The difficulty is that the cognitive and affective domains are not discretely separable in any living human being. Scholars in psychology have separated them for convenience of study, just as physicists find it convenient to think of light sometimes as particle and sometimes as wave. But in fact, light is not either particle or wave, but both. People are not either cognitive or affective, but both. No matter how sophisticated its techniques, measurement that ignores this

What Have They Learned

dual character of human beings is doomed to educational irrelevance, at best.

When data have been collected and researchers want to know what inferences may reasonably be drawn from them, they use statistics. If the data consist of test grades, the inferences drawn from them have no greater value than the grades themselves. Statistical findings based on test scores in science have little value if a valuable measurement is one which helps someone to do something. What can anyone do with these measurements? One might, of course, arrange the children in the order of their test results. Of what real use is this? Will it help to make the children better students, or the testers better teachers? Will such a record be of value to the person who will teach these children the following year? The information extracted from the test results may direct attention to certain children who have not done very well and to others who have scored very high, but most teachers can anticipate these findings. Then what purpose has been served by testing? Will the children who have done badly be taught to be more successful in test-taking? To what end? What is to be done for the children who have done very well? And for those in the middle?

Measurement and statistics, when they are properly employed, help to solve problems. When the problems have been designated, the will forged to seek a solution, measurement can be indispensable. It is pointless to measure for the sake of measuring. Sometimes, however, measuring serves to postpone dealing with a problem while giving the impression of serious endeavor. Countless studies have been made of the ghetto, of the poor, of malnutrition, of neglected children, and of a host of other social problems. Is the failure to feed the hungry really the consequence of an absence of sufficient information? Is the continued presence of economic and social wastelands for millions of Americans the result of inadequate statistical tables? Do the numerical "scores" announced daily from Vietnam bring peace any nearer? The persistence of these problems has not been brought about by an absence of quantitative information. Social and economic problems, like the problem of inadequate science education, result from lack of determination, lack of will, and lack of nerve to find the best means for their solution.

Difficulties in the schools will not be decreased by refinements in testing procedures. They will be exacerbated. Educators will not cultivate a scientifically literate population by perfecting "scientometers" which, when inserted into children's mouths, register the degree of scientific mastery. The need is for recognition of the real, lasting, humane goals for which science is taught. Parents and educators must find the will to struggle against whatever obstacles may stand in their way to provide opportunities for *all* children to reach for these goals as well as they can. They must give greater attention to children, note what is happening to them, and when they have noted, exercise their judgment about what changes will urge the children to reach a little higher.

SELECTED BIBLIOGRAPHY

INDEX

Selected Bibliography

The books here selected for children represent a wide range of scientific activity and are consistent with the point of view expressed in this book. All of them are suitable for children in the elementary school. Those recommended for very young children are marked (P) for primary; those suitable for older children who are already independent readers are marked (I); and those that will appeal to children whose interests and abilities are advanced are marked (A). When a book is appropriate to more than one category, the recommendation is hyphenated—(P-I) or (I-A).

The list of readings for adults consists principally of books but also includes two articles: all are the consequence of profound reflection about children, the school, learning, science and mathematics. Four periodicals which regularly include material related to these interests are listed separately.

For Children

ADLER, IRVING. *The Tools of Science: From Yardstick to Cyclotron*. New York: The John Day Company, Inc., 1958.
The tools of the modern physicist, and how he uses them to observe, analyze, and change parts of his environment. (I-A)

———. *Magic House of Numbers*. New York: The John Day Company, Inc., 1957.
Gives insight into mathematics by means of games, puzzles, and tricks. (I)

———. *Fire in Your Life*. New York: The John Day Company, Inc., 1955.
The role of fire in nature and in human culture. The author discusses various methods of producing fire, of determining temperature, and the technological applications of fire. (I-A)

———. *Things that Spin: From Tops to Atoms*. New York: The John Day Company, Inc., 1960.
The various motions of the earth, moon, sun, stars, and atoms are compared with the motions of a spinning top. (A)

APPIAH, PEGGY. *Tales of an Ashanti Father.* London: André Deutsch, 1967.
Stories from the folk heritage of Ashanti. (P-I)

ASIMOV, ISAAC. *Stars.* New York: Follett Publishing Company, 1968.

———. *Galaxies.* New York: Follett Publishing Company, 1968. (I)
Both these books contain a concise text, excellent illustrations, and a section called "Things to Do." (I)

BENDICK, JEAN. *The Human Senses.* New York: Franklin Watts, Inc., 1968.
Questions are raised which encourage the reader to discover interesting and significant facts about his environment. There are explanations, assisted by diagrams and pictures, of the elementary facts about the physiology of our senses. (I)

———. *Shapes.* New York: Franklin Watts, Inc., 1968.
Introduces the ideas of solid figures, plane figures, line, form, and symmetry. Activities suggested to encourage the child to observe, to classify, and to make inferences. (P)

———. *Space and Time.* New York: Franklin Watts, Inc., 1968.
Space, form, distance and time are related by the use of familiar things and events. (I)

FEIFER, NATHAN. *Adventures in Chemistry.* New York: Sentinel, 1959.
Experiments to reveal some of the properties of water, the small size of molecules, the nature of organic compounds, the variety of ways that calcium occurs, the difference between metals and nonmetals, the nature of food, and even an attempt to imagine the nature of the atom. (A)

MERTENS, ALICE. *Children of the Kalahari.* London: Collins, 1966.
Largely through photographs, the reader is introduced to a way of life very different from his own. Children work, play games, enjoy music, and wait for the return of the hunters. (P-I-A)

MILGROM, HARRY. *Adventures with a Paper Cup.* New York: E. P. Dutton & Co., Inc., 1968.

———. *Adventures with a Straw.* New York: E. P. Dutton & Co., Inc., 1967.

———. *Adventures with a Party Plate.* New York: E. P. Dutton & Co., Inc., 1968.
Activities that require common household articles—children can construct a pinhole camera and a water clock with a cup, whistles with a straw, and can learn the principles of wheels and other lessons in mechanics by making things from party plates. (P-I)

PINE, TILLIE S., and LEVINE, JOSEPH. *The Incas Knew.* New York: McGraw-Hill Book Company, 1968.
Things to do as well as read. Projects are suggested to build models

Selected Bibliography

on Inca patterns and to do experiments employing principles discovered by the Incas. (I-A)

PRINGLE, LAURENCE. *Dinosaurs and Their World*. New York: Harcourt, Brace & World, Inc., 1968.
Careful descriptions illustrated with photographs of field sites, fossils, restorations, paintings, and museum displays. Lists museums where the exhibits can be seen. (I-A)

ROSEN, ELLSWORTH. *Spiders and Spinners*. Boston: Houghton Mifflin Company, 1968.
Combines accurate information, clever drawings, and delightful verse about the structure and life history of spiders. (I)

RUCHLIS, HY. *Clear Thinking*. New York: Harper & Row, Publishers, 1962.
An examination of the underlying principles that govern good thinking patterns, including an introduction to formal logic. Topics considered include the nature of superstition, the connection between language and reasoning, and the techniques employed by advertising to influence thinking. (A)

―――. *Discovering Scientific Method*. New York: Harper & Row, Publishers, 1963.
General principles of scientific methodology, the place of observation, the nature of scientific reasoning, practice in reasoning through the use of picture puzzles, and the importance of measurement and mathematics. (A)

SCHNEIDER, HERMAN, and SCHNEIDER, NINA. *Science Fun with Milk Cartons*. New York: Whittlesey House, 1953.
Instructions and diagrams to build bridges, trucks, trains, boats, elevators, water wheels, windmills, and many other toys from milk cartons. (I-A)

SCHNEIDER, HERMAN. *Everyday Weather and How It Works*. New York: Whittlesey House, 1951.
Explanations of various aspects of weather and instructions for the construction of working weather instruments. (I-A)

―――. *Everyday Machines and How They Work*. New York: Whittlesey House, 1950.
An examination of many household appliances to see what makes them work. Included among these are carpet sweepers, vacuum cleaners, faucets and valves, coffee makers, tools, lights, bells, motors, and many more.

SELSAM, MILLICENT E. *Birth of an Island*. New York: Harper & Row, Publishers, 1959.
How an island evolves—its birth from volcanic eruption, its weathering, and the slow appearance of living things. (P-I)

Teaching Science to Children

———. *See Through the Forest.* New York: Harper & Row, Publishers, 1956.
The ecology of a forest. The various levels of the forest are compared with the floors in a building with an examination of what may be found at each level. (P-I)

———. *Milkweed.* New York: William Morrow & Co., Inc., 1967.
The story of the milkweed by means of photographs and brief text, from the seed, through growth and pollination, to fertilized pods, and the dissemination of seeds again. (P)

SILVERSTEIN, ALVIN, and SILVERSTEIN, VIRGINIA. *A World in a Drop of Water.* New York: Atheneum Publishers, 1969.
Photographs and drawings introduce the world of the very small in a visit to a pond. A simple microscope is required to do the experiments.

SIMON, SEYMOUR. *Animals in Field and Laboratory: Science Projects in Animal Behavior.* New York: McGraw-Hill Book Company, 1968.
Projects for older children to do in the laboratory and in the field which do not require much money, are not technical, and leave room for the children to go off in their own directions. (A)

SLOTE, ALFRED. *Air in Fact and Fancy.* Cleveland: The World Publishing Company, 1968.
Folk tales and facts about thunder, lightning, rain, and storms. It is similar to an earlier book by the same author *The Moon in Fact and Fancy* (World, 1967). (P-I)

SOOTIN, HARRY. *Experiments with Electric Currents.* New York: W. W. Norton & Company, Inc., 1969.
Safe experiments with electricity and magnetism. Materials are inexpensive. (I)

THAPAR, RAJ. *Introducing India.* New York: Asia Publishing House, 1967.
The legends, favorite foods, the role of cotton in India's history, and special days celebrated in various parts of the country. (I-A)

WOLFF, ROBERT J. *Feeling Blue.* New York: Charles Scribner's Sons, 1968.
One of three books for children who are beginning to read. The other two are *Hello Yellow* and *Seeing Red.* Each one tells a story about one of the primary colors through a combination of experiences with abstract art, shapes and colors, the physics of mixing colors, the moods that colors transmit, and some optical illusions. (P)

WYLER, ROSE. *The First Book of Science Experiments.* New York: Franklin Watts, Inc., 1952.
Experiments with air, water, plants, electricity, and light. All of them can be performed with readily available things. (P)

———. *What Makes It Go?* New York: Whittlesey House, 1958.
Simple experiments to understand what enables engines to push, boats to go through water, trains to travel on tracks, and airplanes to fly through the air. (P)

Selected Bibliography

ZIM, HERBERT S., and SKELLY, JAMES R. *Machine Tools.* New York: William Morrow & Co., Inc., 1969.
An introduction to the tools that make mass production possible—the lathe, giant power press, drill press, shaper, milling machine, boring mill, punch press, and others. (A)

For Adults

BERNAL, J. D. "Science Teaching in General Education." In *The Freedom of Necessity.* London: Routledge & Kegan Paul Ltd., 1949. Pp. 135-146.
An essay based on a discussion at Sarah Lawrence College about the part science should play in general education.

BIGGS, EDITH E., and MACLEAN, JAMES R. *Freedom to Learn.* Reading, Mass.: Addison-Wesley Publishing Co., Inc., 1969.
A climate in which children are engaged in their mathematics learning is consistent with the view this book takes about science learning.

BROWN, MARY, and PRECIOUS, NORMAN. *The Integrated Day in the Primary School.* London: Ward Lock Educational, 1968.
The organization of a learning environment characterized by openness. The authors are head teachers in English schools that were among the first to try a new kind of schooling.

BRUNER, JEROME S. *On Knowing.* New York: Atheneum Publishers, 1968.
Ten essays devoted to an exploration of the psychological basis of creating, intuiting, and knowing.

HAWKINS, DAVID. "Messing About in Science," *Science and Children,* 3 (February 1965), 5-9.
Three phases in children's work when they tackle a scientific problem.

HOLT, JOHN. *How Children Fail.* New York: Delta Books, 1965.
The strategies children learn to employ to defend themselves against fear of failure.

———. *The Underachieving School.* New York: Pitman Publishing Corp., 1969.
Essays on the tyranny of testing, the race for college admission, the exclusion of poor children, the absence of activity, and the failure of schools to teach reading.

HONE, ELIZABETH B., JOSEPH, ALEXANDER, and VICTOR, EDWARD. *A Sourcebook for Elementary Science.* New York: Harcourt, Brace & World, Inc., 1962.
Techniques, demonstrations, projects, field trips, and suggestions for instruction in elementary science.

KARPLUS, ROBERT, and THIER, HERBERT D. *A New Look at Elementary School Science.* Chicago: Rand McNally & Co., 1967.

Teaching Science to Children

An account of the Science Curriculum Improvement Study through the summer of 1966.

KUSLAN, LOUIS I., and STONE, A. HARRIS (eds.). *Readings on Teaching Children Science.* Belmont, Calif.: Wadsworth Publishing Co., Inc., 1969.

Thirty-nine selections, principally from the periodical literature, about the goals of science instruction, the nature of science and the child, the curriculum, evaluation, and teacher preparation.

MARSHALL, SYBIL. *An Experiment in Education.* London: Cambridge University Press, 1968.

The "symphonic method" developed by the author over some eighteen years in the primary grades. The focus is art, but the flexibility of the program, which includes both free and traditional teaching, suggests that for someone else science may serve as the focus and be equally successful.

MEARNS, HUGHES. *Creative Power.* New York: Dover Publications, Inc., 1958.

The revival of children's language and literary gifts even after these had been corrupted by poor experience. This book has much to suggest to those who design educational programs, whatever the subject.

MORRISON, PHILIP. "Experimenters in the Schoolroom," *Science,* 138 (December 21, 1962), 1307-1310.

A description of the efforts of Elementary Science Study to develop experiences in science for children.

SCHMIDT, VICTOR E., and ROCKASTLE, VERNE N. *Teaching Science with Everyday Things.* New York: McGraw-Hill Book Company, 1968.

Suggestions for first hand experiences with real things.

VICTOR, EDWARD, and LERNER, MARJORIE S. (eds.). *Readings in Science Education for the Elementary School.* New York: The Macmillan Company, 1967.

Articles dealing with current thinking, especially with ideas that have been developed since 1960, in elementary science.

Journals and Periodicals

Environment. An official publication of Scientists' Institute for Public Information, published by the Committee for Environmental Information, 438 N. Skinker Boulevard, St. Louis, Mo. 63130.

Information about the effects of technology on the environment.

Natural History. The journal of the American Museum of Natural History, Central Park West at 79th Street, New York, N.Y. 10024.

Articles on a variety of topics in the bio-social field. Each issue also contains a sky map for the month and outstanding features to note.

Selected Bibliography

Science and Children. The journal of the National Science Teachers Association, 1201 16th Street, N.W., Washington, D.C. 20036.

The professional journal devoted to science in the elementary school. Articles deal with unusually successful lessons, curriculum problems, articulation with other subjects, and other matters of interest to teachers.

Science Books: A Quarterly Review. HILARY J. DEASON (ed.). American Association for the Advancement of Science, 1515 Massachusetts Avenue, N. W. Washington, D.C.

Reviews books for children and adults in the pure and applied sciences. Books are graded according to school level, and they are marked according to their quality.

Index

activities, choice of, 34-36
activity, 91
Agassiz, Jean Louis Rodolphe, 107
Alcindor, Lew, 89
Alcmaeon, 80
alcohol molecules, 46
algae, 38
American culture, 27
anabolism, 95
animals, caring for, 30
animals, large, 88-89
answer syndrome, 69
anthropology, 102
anti-authoritarian tradition, 5
ants, 38-39, 77
Appiah, Peggy, 67
art, prehistoric, 92-94
arts, visual, 101
Auden, W. H. (quoted), 120
Avogadro, Amedeo, 108

balloons, 32-34
Banneker, Benjamin, 11
Barnett, S. A., 39
Barr, Alfred H., Jr., 96
behavior patterns, 54-55
benzene molecules, 47-48
Berlese funnel, 37
Bernal, J. D., 30
Bernard, Claude, 107
Berzelius, Jöns Jacob, 54, 108
Beveridge, W. I. B. (quoted), 63
Biggs, Edith, 90
Bitter, Francis, 84
Boaz, Franz (quoted), 99
Bohr, Niels, 31, 48
Boys, C. V., 25, 36
bracket fungus, 38
Bridgman, Percy W., 45
Bruno, Giordano, 5

bulletin board, 73
Buridan, Jean, 106-107

Calandra, Alexander, 126
Cannizzaro, Stanislao, 49, 108
Carson, Rachel, 22
Carver, George Washington, 11
catabolism, 95
Cavendish, Henry, 50, 108
Cavendish laboratory, 51-52
cave paintings, 92-94
cephenomyia, 77
Cervantes Saavedra, Miguel de, 44
Cherokee Indians, 67
Chukovsky, Kornei, 66
Clark, Kenneth B. (quoted), 100
Coffin, Robert P. Tristram (quoted), 38
Colbert, Edwin H. (quoted), 41
commitment, lack of, 15
Compton, Arthur, 7
Copernican model, 13-14
Copernicus, 5, 6, 48
Craig, Gerald S., v
creative imagination, 64
Crick, Francis H. C., 48
Crow Indians, 67
crystals, 87
Cullen, Countee (quoted), 99-100
cultures, nonliterate, 26-27
Cummings, E. E. (quoted), 66
Curie, Marie and Pierre, 43

Dalton, John, 108
dandelions, 39
Darwin, Charles, 29, 39, 54, 64-65, 102, 107
de Beer, Gavin, 107
demonstration techniques, 71-72

Dewey, John, v, 30
Dickinson, Emily (quoted), 40
dinosaurs, 88
Dobuan islanders, 68
Drew, Charles R., 11
Dubois, Marie Eugène François Thomas, 110
Dubos, René, 64

earth, gravity of, 77
earthworms, 39
Eddington, Arthur Stanley, 81
Einstein, Albert, 7, 11, 31, 43, 81, 95
Eiseley, Loren (quoted), 64-65
Eiss, Albert F., vi
elephants, 88
Eliot, George (quoted), 67
English language, 20
environment, 75-77
enzymes, 110
equipment, variety of, 58
error, 106
esthetic experience, 92
ether, 11
experiments, 81-87
experts, contributions of, 59
exploration, free, 28
explorations, 118

failure, 106
family experience, 28
Faraday, Michael, 80
Farrington, Benjamin, 79
ferns, liverworts, and mosses, 38
Feynman, Richard, 20-21, 85
Fourier, Jean Baptiste Joseph, 109
Frankland, Sir Edward, 49
Fraunhofer, Joseph von, 109
Fresnel, Augustin Jean, 7
Froebel, Friedrich, 30
Fukuda, Soichi, 52

Gallen, 80
Galileo, 6, 46, 81, 107
game metaphor, 44-45
games, 28, 42 ff
gases, volume of, 108
genes, 66
germ plasm theory, 65
Gesell, Arnold, 30
"giant project," 88-89
Gibbs, Josiah Willard, 11
Gogh, Vincent van, 43
Greenaway, Kate (quoted), 24

Haldane, J. B. S., 36, 88
Handel, Georg Friedrich, 44
Harbeck, Mary Blatt, vi
Harvey, William, 8, 17 (quoted)
Haüy, Père, 87
Hawkins, David, 22
heat, 8, 109
Henry, Joseph, 11, 84, 85
Herschel, William, 107
Hinton, William A., 11
Hoffmann, Banesh, 126
Holub, Miroslav (quoted), 98-99, 105
Hooke, Robert, 50, 69
hormones, 52-53
human races, creation legends of, 67-68
Hutton, James, 8
Huygens, Christian, 7, 110, 123

ignorance, admission of, 19
individual interest, 32
insects, 39
instruction, traditional, 58-59
inventiveness, 78-79
Isaacs, Susan, 30
isomers, 46-47

Index

Jaffe, Bernard, 69
Jameson, Robert, 8
Java man, 110
Jenner, Edward, 109
Jivaro Indians, 26
Joachim, Georges, 6
Job, Book of (quoted), 62
Joule, James Prescott, 109
Julian, Percy L., 11
Just, Ernest Everett, 11

Kalahari boy, 18
Kaplan, Abraham, 45
Karlson, Peter, 52
Karplus, Robert, 22
Keats, John (quoted), 92
Kekulé, Friedrich August, 46-48
Kelvin, William Thomson, Lord (quoted), 102; 109, 127
Kepler, Johannes (quoted), 45-46
Kettle, Arnold, 97
Kieran, John, 21
Kline, Morris (quoted), 90-91
Kluckhohn, Clyde, 26
Kopéc, Stefan, 52

laboratory, outdoor, 36
Lagrange, Joseph Louis, 109
Lamarck, Jean Baptiste Pierre Antoine de Monet, 64
Langmuir, Irving, 69
Laplace, Pierre Simon, 109
Lapps, 27
laser demonstration, 59-60
La Varre, W., 37
Lavoisier, Antoine Laurent, 11
learning:
  climate, 14-15, 17, 22-23
  environment, 70
  evidence, 121-125
  experience, 16

Le Bel, Joseph Achille, 48
Legendre, Adrian Marie, 109
legends, 67
Lewis, Gilbert Newton, 48
lichen, 38
light:
  electromagnetic theory, 7
  nature of, 8
  wave model of, 7
Lissajou figures, 35
living things, learning about, 36
Lutz, Frank E., 21
Lyell, Charles, 8, 64, 102, 107
Lyonet, Pierre, 52

Macdonald, D. K. C., 69
Maclean, James, 90
magnetism, 86-87
Makarenko, A. S., 30
man, uniqueness of, 76-78
manual work, 79
Marx, Karl, 83
mathematics, 16, 66, 90
mathematics, "new," 16
Matzeliger, Jan, 11
Maxwell, James Clerk, 7, 69
Mearns, Hughes (quoted), 3
measurement, 127, 128, 129
Mechnikov, Ilya (quoted), 63; 64
Medawar, P. B., 31, 81, 84
Mendel, Gregor Johann, 108
Mendeléev, Dmitri Ivanovich, 29
metabolism, 95
Michelangelo, 43
Michelson, Albert Abraham, 11
Miescher, Friedrich, 51
millipede, 37
Milton, John, 97
mirrors, 85-86
mistakes, 106-107, 110-114
models, 48-49, 53-54
Montessori, Maria, 30

143

Teaching Science to Children

moon landing, 70-71
Moore, Henry, 24, 94
Morrison, Philip, 98
museum trip, 60
mushrooms, 38
music, 101
mystery boxes, 55

Nägeli, Karl W., von, 108
names, 115-116
National Science Foundation, v
Navajo Indians, 26
Needham, John Turberville, 9
Needham, Joseph, 95
Neill, A. S., 30
neptunists, 8
Newton, Isaac, 7, 29, 107
New York City, field trips in, 21
"nodding committees," 50
Northup, John Howard, 110

observation, powers of, 19
Ojibwa women, 27

Pasteur, Louis, 9, 46, 63-64
   (quoted), 109
Pauling, Linus, 46, 48, 96, 97
Pavlov, Ivan Petrovich, 54
pendulum, 34-35, 81, 123, 124
pendulum, Foucault, 114
photoelectric effect, 7
Piaget, Jean, 30
Picasso, Pablo, 96, 97
pit vipers, senses of, 76
Planck, Max, 7
planetarium, 60, 72-73
plants and animals, 36-37
play:
   episodes, 40-41
   and imagination, 28
   instructive, 25-26
   useful activity, 29-31

Playfair, John, 8
poetry, 101
Priestley, Joseph, 10-11, 108
progress, child's, 120-121
protozoans, digestion in, 63
Ptolemy, 6, 48
puffball, 28
Puritan tradition, 25
puzzles, 45, 55, 73, 125
Pygmies, 18, 27
Pythagorean theorem, 90

questions:
   children's, 66-67
   essential, 12-13
   gift of, 69
   and problems, 62 ff
   variety, 74

raccoon, 77
random activities, 56
Raynal, Maurice, 96
Redi, Francesco, 8, 51
relativity, 11, 95
Rheticus, 6
Riemann, Georg F., 91
Rillieux, Norbert, 11
road trips, 60-61
Rockefeller, Nelson, 24
Ross, Arthur, 127
Rutherford, Ernest, 110

Sagan, Carl (quoted), 101
Sandburg, Carl (quoted), 6
Saxena, Kailash, 52
senses, dulled, 19
school, 49-50
schooling, meaning of, 13
science-mindedness, 18, 22
Scientific American, 50
scientists, biographies of, 68

Index

sensory activity, 76-77
shadow game, 111-112
Shelley, Percy Bysshe, 97
Sláma, Karel, 52
Smith, William, 8
snail, 37
soap bubbles, 36
Socrates, 69-70
solar system, 6
sound, 4
sowbug, 37
Spallanzani, Lazzaro, 9
Spender, Stephen, 43
spiders, 38
spontaneous generation, 8, 9, 63-64
springtails, 37
Sputnik I, v
statistics, 129
Suchman, J. Richard (quoted), 69
Sumner, James Batcheller, 110
surgery, 80

task cards, 55-56
teacher, as laboratory director, 56-57
teaching, successful, 103
television-watching, 18
testing, 126, 127
Thapar, Raj (quoted), 67
theory and practice, 10
Thomson, Joseph John, 51-52
towns, growth of, 94
toys, 18
trypanosome family, 76
Tsiolkovsky, Konstantin Eduardovich, 29, 54
Turnbull, Colin, 27
Twain, Mark, 24

universe, geocentric, 6

values, 2
Van Doren, Mark (quoted), 100, 116
van't Hoff, Jacobus Hendricus, 48
Vesalius, Andreas, 79
Virchow, Rudolph, 108
Vishnu, 67
vocabulary, 18
vulcanists, 8

Waddington, C. H., 78
Wald, George, 67
water displacement, 90
water play, 30
Watson, James D. (quoted) 24-25, 97; 48
Webster, David, 22
Weigle, Jean, 97
Weismann, August F. L., 65
Werner, Abraham Gottlob, 8
whales, senses of, 76
Whitman, Walt (quoted), 42
Wigglesworth, Vincent, 52
Williams, Carroll, 52
Williamson, Alexander W., 49
Willstätter, Richard, 110
withdrawal symptom, 56
Wollaston, William Hyde, 109
wooden blocks, 30
Woods, Granville T., 11
words, 115-116
work experience, 42-43
work and play, 24-25

Xenophon, 79
X rays, 7

Yin and Yang, 95
Young, Thomas, 7

Zalkind, Semyon, 63
zoo, 60

About the Author

Lazer Goldberg was an assistant professor in the School of Education at Hofstra University in Hempstead, New York, and consultant to The Little Red School House in New York City. He received both his M.A. (1947) and his Ed.D. (1966) from Teachers College, Columbia University. Dr. Goldberg taught children from age four through high school and taught adults in undergraduate and graduate school. He gave lectures to Head Start administrators in New England and New Jersey and summer workshops at the Montessori Teacher Training Program (Fairleigh Dickinson University, Teaneck, New Jersey). Dr. Goldberg was the author of *Chemistry with Safe Experiments You Can Do* and contributed to the magazine *Science and Children*.

Acknowledgments

The following poems or excerpts are reprinted by permission of the publisher or other source indicated:

"Slow Miss Mandy" from *Creative Power* by Hughes Mearns. Dover Publications, Inc., New York.

The People, Yes by Carl Sandburg (Copyright, 1936, by Harcourt, Brace & World, Inc.; Copyright, 1964, by Carl Sandburg). Harcourt, Brace & World, Inc., New York.

"School is Over" from *Under the Window* by Kate Greenaway. Frederick Warne and Company, New York.

"The Spider" from *Collected Poems* by Robert P. Tristam Coffin (Copyright 1929 by The Macmillan Company, renewed 1957 by Robert P. Tristam Coffin). The Macmillan Company, New York.

"It troubled me as once I was" (#150) from *Bolts of Melody, New Poems of Emily Dickinson*, edited by Mabel Loomis Todd and Millicent Todd Bingham (Copyright, 1945 by The Trustees of Amherst College). Harper & Row, Publishers, New York.

"The Lesson" and "Inventions" from *Selected Poems* by Miroslav Holub, translated by Ian F. G. Milner and George Theiner, published by Penguin Books, Ltd. By permission of Ian F. G. Milner.

"If They Spoke" from *Spring Birth* and "The God of Galaxies" by Mark Van Doren, which are both included in *Collected and New Poems* (Copyright © 1963 by Mark Van Doren). Hill and Wang, Inc., New York.

"The Unknown Citizen" from *The Collected Poetry of W. H. Auden* (Copyright 1940 and renewed 1968 by W. H. Auden). Random House, Inc., New York, and Faber and Faber Ltd., London.

"Incident" from *On These I Stand* by Countee Cullen (Copyright 1925 by Harper & Brothers; renewed, 1953 by Ida M. Cullen). Harper & Row, Publishers.

A CATALOG OF SELECTED
DOVER BOOKS
IN ALL FIELDS OF INTEREST

A CATALOG OF SELECTED DOVER BOOKS IN ALL FIELDS OF INTEREST

CONCERNING THE SPIRITUAL IN ART, Wassily Kandinsky. Pioneering work by father of abstract art. Thoughts on color theory, nature of art. Analysis of earlier masters. 12 illustrations. 80pp. of text. 5⅜ × 8½. 23411-8 Pa. $3.95

ANIMALS: 1,419 Copyright-Free Illustrations of Mammals, Birds, Fish, Insects, etc., Jim Harter (ed.). Clear wood engravings present, in extremely lifelike poses, over 1,000 species of animals. One of the most extensive pictorial sourcebooks of its kind. Captions. Index. 284pp. 9 × 12. 23766-4 Pa. $12.95

CELTIC ART: The Methods of Construction, George Bain. Simple geometric techniques for making Celtic interlacements, spirals, Kells-type initials, animals, humans, etc. Over 500 illustrations. 160pp. 9 × 12. (USO) 22923-8 Pa. $9.95

AN ATLAS OF ANATOMY FOR ARTISTS, Fritz Schider. Most thorough reference work on art anatomy in the world. Hundreds of illustrations, including selections from works by Vesalius, Leonardo, Goya, Ingres, Michelangelo, others. 593 illustrations. 192pp. 7⅛ × 10¼. 20241-0 Pa. $9.95

CELTIC HAND STROKE-BY-STROKE (Irish Half-Uncial from "The Book of Kells"): An Arthur Baker Calligraphy Manual, Arthur Baker. Complete guide to creating each letter of the alphabet in distinctive Celtic manner. Covers hand position, strokes, pens, inks, paper, more. Illustrated. 48pp. 8¼ × 11. 24336-2 Pa. $3.95

EASY ORIGAMI, John Montroll. Charming collection of 32 projects (hat, cup, pelican, piano, swan, many more) specially designed for the novice origami hobbyist. Clearly illustrated easy-to-follow instructions insure that even beginning papercrafters will achieve successful results. 48pp. 8¼ × 11. 27298-2 Pa. $2.95

THE COMPLETE BOOK OF BIRDHOUSE CONSTRUCTION FOR WOODWORKERS, Scott D. Campbell. Detailed instructions, illustrations, tables. Also data on bird habitat and instinct patterns. Bibliography. 3 tables. 63 illustrations in 15 figures. 48pp. 5¼ × 8½. 24407-5 Pa. $1.95

BLOOMINGDALE'S ILLUSTRATED 1886 CATALOG: Fashions, Dry Goods and Housewares, Bloomingdale Brothers. Famed merchants' extremely rare catalog depicting about 1,700 products: clothing, housewares, firearms, dry goods, jewelry, more. Invaluable for dating, identifying vintage items. Also, copyright-free graphics for artists, designers. Co-published with Henry Ford Museum & Greenfield Village. 160pp. 8¼ × 11. 25780-0 Pa. $9.95

HISTORIC COSTUME IN PICTURES, Braun & Schneider. Over 1,450 costumed figures in clearly detailed engravings—from dawn of civilization to end of 19th century. Captions. Many folk costumes. 256pp. 8⅜ × 11¼. 23150-X Pa. $11.95

CATALOG OF DOVER BOOKS

STICKLEY CRAFTSMAN FURNITURE CATALOGS, Gustav Stickley and L. & J. G. Stickley. Beautiful, functional furniture in two authentic catalogs from 1910. 594 illustrations, including 277 photos, show settles, rockers, armchairs, reclining chairs, bookcases, desks, tables. 183pp. 6½ × 9¼. 23838-5 Pa. $9.95

AMERICAN LOCOMOTIVES IN HISTORIC PHOTOGRAPHS: 1858 to 1949, Ron Ziel (ed.). A rare collection of 126 meticulously detailed official photographs, called "builder portraits," of American locomotives that majestically chronicle the rise of steam locomotive power in America. Introduction. Detailed captions. xi + 129pp. 9 × 12. 27393-8 Pa. $12.95

AMERICA'S LIGHTHOUSES: An Illustrated History, Francis Ross Holland, Jr. Delightfully written, profusely illustrated fact-filled survey of over 200 American lighthouses since 1716. History, anecdotes, technological advances, more. 240pp. 8 × 10¾. 25576-X Pa. $11.95

TOWARDS A NEW ARCHITECTURE, Le Corbusier. Pioneering manifesto by founder of "International School." Technical and aesthetic theories, views of industry, economics, relation of form to function, "mass-production split" and much more. Profusely illustrated. 320pp. 6⅛ × 9¼. (USO) 25023-7 Pa. $9.95

HOW THE OTHER HALF LIVES, Jacob Riis. Famous journalistic record, exposing poverty and degradation of New York slums around 1900, by major social reformer. 100 striking and influential photographs. 233pp. 10 × 7⅞.
22012-5 Pa $10.95

FRUIT KEY AND TWIG KEY TO TREES AND SHRUBS, William M. Harlow. One of the handiest and most widely used identification aids. Fruit key covers 120 deciduous and evergreen species; twig key 160 deciduous species. Easily used. Over 300 photographs. 126pp. 5⅜ × 8½. 20511-8 Pa. $3.95

COMMON BIRD SONGS, Dr. Donald J. Borror. Songs of 60 most common U.S. birds: robins, sparrows, cardinals, bluejays, finches, more—arranged in order of increasing complexity. Up to 9 variations of songs of each species.
Cassette and manual 99911-4 $8.95

ORCHIDS AS HOUSE PLANTS, Rebecca Tyson Northen. Grow cattleyas and many other kinds of orchids—in a window, in a case, or under artificial light. 63 illustrations. 148pp. 5⅜ × 8½. 23261-1 Pa. $4.95

MONSTER MAZES, Dave Phillips. Masterful mazes at four levels of difficulty. Avoid deadly perils and evil creatures to find magical treasures. Solutions for all 32 exciting illustrated puzzles. 48pp. 8¼ × 11. 26005-4 Pa. $2.95

MOZART'S DON GIOVANNI (DOVER OPERA LIBRETTO SERIES), Wolfgang Amadeus Mozart. Introduced and translated by Ellen H. Bleiler. Standard Italian libretto, with complete English translation. Convenient and thoroughly portable—an ideal companion for reading along with a recording or the performance itself. Introduction. List of characters. Plot summary. 121pp. 5¼ × 8½.
24944-1 Pa. $2.95

TECHNICAL MANUAL AND DICTIONARY OF CLASSICAL BALLET, Gail Grant. Defines, explains, comments on steps, movements, poses and concepts. 15-page pictorial section. Basic book for student, viewer. 127pp. 5⅜ × 8½.
21843-0 Pa. $4.95

BRASS INSTRUMENTS: Their History and Development, Anthony Baines. Authoritative, updated survey of the evolution of trumpets, trombones, bugles, cornets, French horns, tubas and other brass wind instruments. Over 140 illustrations and 48 music examples. Corrected and updated by author. New preface. Bibliography. 320pp. 5⅜ × 8½. 27574-4 Pa. $9.95

HOLLYWOOD GLAMOR PORTRAITS, John Kobal (ed.). 145 photos from 1926–49. Harlow, Gable, Bogart, Bacall; 94 stars in all. Full background on photographers, technical aspects. 160pp. 8⅜ × 11¼. 23352-9 Pa. $11.95

MAX AND MORITZ, Wilhelm Busch. Great humor classic in both German and English. Also 10 other works: "Cat and Mouse," "Plisch and Plumm," etc. 216pp. 5⅜ × 8½. 20181-3 Pa. $5.95

THE RAVEN AND OTHER FAVORITE POEMS, Edgar Allan Poe. Over 40 of the author's most memorable poems: "The Bells," "Ulalume," "Israfel," "To Helen," "The Conqueror Worm," "Eldorado," "Annabel Lee," many more. Alphabetic lists of titles and first lines. 64pp. $5\frac{3}{16}$ × 8¼. 26685-0 Pa. $1.00

SEVEN SCIENCE FICTION NOVELS, H. G. Wells. The standard collection of the great novels. Complete, unabridged. First Men in the Moon, Island of Dr. Moreau, War of the Worlds, Food of the Gods, Invisible Man, Time Machine, In the Days of the Comet. Total of 1,015pp. 5⅜ × 8½. (USO) 20264-X Clothbd. $29.95

AMULETS AND SUPERSTITIONS, E. A. Wallis Budge. Comprehensive discourse on origin, powers of amulets in many ancient cultures: Arab, Persian, Babylonian, Assyrian, Egyptian, Gnostic, Hebrew, Phoenician, Syriac, etc. Covers cross, swastika, crucifix, seals, rings, stones, etc. 584pp. 5⅜ × 8½. 23573-4 Pa. $12.95

RUSSIAN STORIES/PYCCKNE PACCKA3bl: A Dual-Language Book, edited by Gleb Struve. Twelve tales by such masters as Chekhov, Tolstoy, Dostoevsky, Pushkin, others. Excellent word-for-word English translations on facing pages, plus teaching and study aids, Russian/English vocabulary, biographical/critical introductions, more. 416pp. 5⅜ × 8½. 26244-8 Pa. $8.95

PHILADELPHIA THEN AND NOW: 60 Sites Photographed in the Past and Present, Kenneth Finkel and Susan Oyama. Rare photographs of City Hall, Logan Square, Independence Hall, Betsy Ross House, other landmarks juxtaposed with contemporary views. Captures changing face of historic city. Introduction. Captions. 128pp. 8¼ × 11. 25790-8 Pa. $9.95

AIA ARCHITECTURAL GUIDE TO NASSAU AND SUFFOLK COUNTIES, LONG ISLAND, The American Institute of Architects, Long Island Chapter, and the Society for the Preservation of Long Island Antiquities. Comprehensive, well-researched and generously illustrated volume brings to life over three centuries of Long Island's great architectural heritage. More than 240 photographs with authoritative, extensively detailed captions. 176pp. 8¼ × 11. 26946-9 Pa. $14.95

NORTH AMERICAN INDIAN LIFE: Customs and Traditions of 23 Tribes, Elsie Clews Parsons (ed.). 27 fictionalized essays by noted anthropologists examine religion, customs, government, additional facets of life among the Winnebago, Crow, Zuni, Eskimo, other tribes. 480pp. 6⅛ × 9¼. 27377-6 Pa. $10.95

CATALOG OF DOVER BOOKS

FRANK LLOYD WRIGHT'S HOLLYHOCK HOUSE, Donald Hoffmann. Lavishly illustrated, carefully documented study of one of Wright's most controversial residential designs. Over 120 photographs, floor plans, elevations, etc. Detailed perceptive text by noted Wright scholar. Index. 128pp. 9¼ × 10¾.
27133-1 Pa. $11.95

THE MALE AND FEMALE FIGURE IN MOTION: 60 Classic Photographic Sequences, Eadweard Muybridge. 60 true-action photographs of men and women walking, running, climbing, bending, turning, etc., reproduced from rare 19th-century masterpiece. vi + 121pp. 9 × 12.
24745-7 Pa. $10.95

1001 QUESTIONS ANSWERED ABOUT THE SEASHORE, N. J. Berrill and Jacquelyn Berrill. Queries answered about dolphins, sea snails, sponges, starfish, fishes, shore birds, many others. Covers appearance, breeding, growth, feeding, much more. 305pp. 5¼ × 8¼.
23366-9 Pa. $7.95

GUIDE TO OWL WATCHING IN NORTH AMERICA, Donald S. Heintzelman. Superb guide offers complete data and descriptions of 19 species: barn owl, screech owl, snowy owl, many more. Expert coverage of owl-watching equipment, conservation, migrations and invasions, etc. Guide to observing sites. 84 illustrations. xiii + 193pp. 5⅜ × 8½.
27344-X Pa. $8.95

MEDICINAL AND OTHER USES OF NORTH AMERICAN PLANTS: A Historical Survey with Special Reference to the Eastern Indian Tribes, Charlotte Erichsen-Brown. Chronological historical citations document 500 years of usage of plants, trees, shrubs native to eastern Canada, northeastern U.S. Also complete identifying information. 343 illustrations. 544pp. 6½ × 9¼.
25951-X Pa. $12.95

STORYBOOK MAZES, Dave Phillips. 23 stories and mazes on two-page spreads: Wizard of Oz, Treasure Island, Robin Hood, etc. Solutions. 64pp. 8¼ × 11.
23628-5 Pa. $2.95

NEGRO FOLK MUSIC, U.S.A., Harold Courlander. Noted folklorist's scholarly yet readable analysis of rich and varied musical tradition. Includes authentic versions of over 40 folk songs. Valuable bibliography and discography. xi + 324pp. 5⅜ × 8½.
27350-4 Pa. $7.95

MOVIE-STAR PORTRAITS OF THE FORTIES, John Kobal (ed.). 163 glamor, studio photos of 106 stars of the 1940s: Rita Hayworth, Ava Gardner, Marlon Brando, Clark Gable, many more. 176pp. 8⅜ × 11¼.
23546-7 Pa. $11.95

BENCHLEY LOST AND FOUND, Robert Benchley. Finest humor from early 30s, about pet peeves, child psychologists, post office and others. Mostly unavailable elsewhere. 73 illustrations by Peter Arno and others. 183pp. 5⅜ × 8½.
22410-4 Pa. $5.95

YEKL and THE IMPORTED BRIDEGROOM AND OTHER STORIES OF YIDDISH NEW YORK, Abraham Cahan. Film Hester Street based on Yekl (1896). Novel, other stories among first about Jewish immigrants on N.Y.'s East Side. 240pp. 5⅜ × 8½.
22427-9 Pa. $6.95

SELECTED POEMS, Walt Whitman. Generous sampling from *Leaves of Grass*. Twenty-four poems include "I Hear America Singing," "Song of the Open Road," "I Sing the Body Electric," "When Lilacs Last in the Dooryard Bloom'd," "O Captain! My Captain!"—all reprinted from an authoritative edition. Lists of titles and first lines. 128pp. 5³/₁₆ × 8¼.
26878-0 Pa. $1.00

CATALOG OF DOVER BOOKS

THE BEST TALES OF HOFFMANN, E. T. A. Hoffmann. 10 of Hoffmann's most important stories: "Nutcracker and the King of Mice," "The Golden Flowerpot," etc. 458pp. 5⅜ × 8½. 21793-0 Pa. $8.95

FROM FETISH TO GOD IN ANCIENT EGYPT, E. A. Wallis Budge. Rich detailed survey of Egyptian conception of "God" and gods, magic, cult of animals, Osiris, more. Also, superb English translations of hymns and legends. 240 illustrations. 545pp. 5⅜ × 8½. 25803-3 Pa. $11.95

FRENCH STORIES/CONTES FRANÇAIS: A Dual-Language Book, Wallace Fowlie. Ten stories by French masters, Voltaire to Camus: "Micromegas" by Voltaire; "The Atheist's Mass" by Balzac; "Minuet" by de Maupassant; "The Guest" by Camus, six more. Excellent English translations on facing pages. Also French-English vocabulary list, exercises, more. 352pp. 5⅜ × 8½. 26443-2 Pa. $8.95

CHICAGO AT THE TURN OF THE CENTURY IN PHOTOGRAPHS: 122 Historic Views from the Collections of the Chicago Historical Society, Larry A. Viskochil. Rare large-format prints offer detailed views of City Hall, State Street, the Loop, Hull House, Union Station, many other landmarks, circa 1904-1913. Introduction. Captions. Maps. 144pp. 9⅜ × 12¼. 24656-6 Pa. $12.95

OLD BROOKLYN IN EARLY PHOTOGRAPHS, 1865-1929, William Lee Younger. Luna Park, Gravesend race track, construction of Grand Army Plaza, moving of Hotel Brighton, etc. 157 previously unpublished photographs. 165pp. 8⅞ × 11¾. 23587-4 Pa. $13.95

THE MYTHS OF THE NORTH AMERICAN INDIANS, Lewis Spence. Rich anthology of the myths and legends of the Algonquins, Iroquois, Pawnees and Sioux, prefaced by an extensive historical and ethnological commentary. 36 illustrations. 480pp. 5⅜ × 8½. 25967-6 Pa. $8.95

AN ENCYCLOPEDIA OF BATTLES: Accounts of Over 1,560 Battles from 1479 B.C. to the Present, David Eggenberger. Essential details of every major battle in recorded history from the first battle of Megiddo in 1479 B.C. to Grenada in 1984. List of Battle Maps. New Appendix covering the years 1967-1984. Index. 99 illustrations. 544pp. 6½ × 9¼. 24913-1 Pa. $14.95

SAILING ALONE AROUND THE WORLD, Captain Joshua Slocum. First man to sail around the world, alone, in small boat. One of great feats of seamanship told in delightful manner. 67 illustrations. 294pp. 5⅜ × 8½. 20326-3 Pa. $5.95

ANARCHISM AND OTHER ESSAYS, Emma Goldman. Powerful, penetrating, prophetic essays on direct action, role of minorities, prison reform, puritan hypocrisy, violence, etc. 271pp. 5⅜ × 8½. 22484-8 Pa. $5.95

MYTHS OF THE HINDUS AND BUDDHISTS, Ananda K. Coomaraswamy and Sister Nivedita. Great stories of the epics; deeds of Krishna, Shiva, taken from puranas, Vedas, folk tales; etc. 32 illustrations. 400pp. 5⅜ × 8½. 21759-0 Pa. $9.95

BEYOND PSYCHOLOGY, Otto Rank. Fear of death, desire of immortality, nature of sexuality, social organization, creativity, according to Rankian system. 291pp. 5⅜ × 8½. 20485-5 Pa. $8.95

A THEOLOGICO-POLITICAL TREATISE, Benedict Spinoza. Also contains unfinished Political Treatise. Great classic on religious liberty, theory of government on common consent. R. Elwes translation. Total of 421pp. 5⅜ × 8½. 20249-6 Pa. $8.95

CATALOG OF DOVER BOOKS

MY BONDAGE AND MY FREEDOM, Frederick Douglass. Born a slave, Douglass became outspoken force in antislavery movement. The best of Douglass' autobiographies. Graphic description of slave life. 464pp. 5⅜ × 8½. 22457-0 Pa. $8.95

FOLLOWING THE EQUATOR: A Journey Around the World, Mark Twain. Fascinating humorous account of 1897 voyage to Hawaii, Australia, India, New Zealand, etc. Ironic, bemused reports on peoples, customs, climate, flora and fauna, politics, much more. 197 illustrations. 720pp. 5⅜ × 8½. 26113-1 Pa. $15.95

THE PEOPLE CALLED SHAKERS, Edward D. Andrews. Definitive study of Shakers: origins, beliefs, practices, dances, social organization, furniture and crafts, etc. 33 illustrations. 351pp. 5⅜ × 8½. 21081-2 Pa. $8.95

THE MYTHS OF GREECE AND ROME, H. A. Guerber. A classic of mythology, generously illustrated, long prized for its simple, graphic, accurate retelling of the principal myths of Greece and Rome, and for its commentary on their origins and significance. With 64 illustrations by Michelangelo, Raphael, Titian, Rubens, Canova, Bernini and others. 480pp. 5⅜ × 8½. 27584-1 Pa. $9.95

PSYCHOLOGY OF MUSIC, Carl E. Seashore. Classic work discusses music as a medium from psychological viewpoint. Clear treatment of physical acoustics, auditory apparatus, sound perception, development of musical skills, nature of musical feeling, host of other topics. 88 figures. 408pp. 5⅜ × 8½. 21851-1 Pa. $9.95

THE PHILOSOPHY OF HISTORY, Georg W. Hegel. Great classic of Western thought develops concept that history is not chance but rational process, the evolution of freedom. 457pp. 5⅜ × 8½. 20112-0 Pa. $9.95

THE BOOK OF TEA, Kakuzo Okakura. Minor classic of the Orient: entertaining, charming explanation, interpretation of traditional Japanese culture in terms of tea ceremony. 94pp. 5⅜ × 8½. 20070-1 Pa. $3.95

LIFE IN ANCIENT EGYPT, Adolf Erman. Fullest, most thorough, detailed older account with much not in more recent books, domestic life, religion, magic, medicine, commerce, much more. Many illustrations reproduce tomb paintings, carvings, hieroglyphs, etc. 597pp. 5⅜ × 8½. 22632-8 Pa. $10.95

SUNDIALS, Their Theory and Construction, Albert Waugh. Far and away the best, most thorough coverage of ideas, mathematics concerned, types, construction, adjusting anywhere. Simple, nontechnical treatment allows even children to build several of these dials. Over 100 illustrations. 230pp. 5⅜ × 8½. 22947-5 Pa. $7.95

DYNAMICS OF FLUIDS IN POROUS MEDIA, Jacob Bear. For advanced students of ground water hydrology, soil mechanics and physics, drainage and irrigation engineering, and more. 335 illustrations. Exercises, with answers. 784pp. 6⅛ × 9¼. 65675-6 Pa. $19.95

SONGS OF EXPERIENCE: Facsimile Reproduction with 26 Plates in Full Color, William Blake. 26 full-color plates from a rare 1826 edition. Includes "The Tyger," "London," "Holy Thursday," and other poems. Printed text of poems. 48pp. 5¼ × 7. 24636-1 Pa. $4.95

OLD-TIME VIGNETTES IN FULL COLOR, Carol Belanger Grafton (ed.). Over 390 charming, often sentimental illustrations, selected from archives of Victorian graphics—pretty women posing, children playing, food, flowers, kittens and puppies, smiling cherubs, birds and butterflies, much more. All copyright-free. 48pp. 9¼ × 12¼. 27269-9 Pa. $5.95

CATALOG OF DOVER BOOKS

PERSPECTIVE FOR ARTISTS, Rex Vicat Cole. Depth, perspective of sky and sea, shadows, much more, not usually covered. 391 diagrams, 81 reproductions of drawings and paintings. 279pp. 5⅜ × 8½. 22487-2 Pa. $6.95

DRAWING THE LIVING FIGURE, Joseph Sheppard. Innovative approach to artistic anatomy focuses on specifics of surface anatomy, rather than muscles and bones. Over 170 drawings of live models in front, back and side views, and in widely varying poses. Accompanying diagrams. 177 illustrations. Introduction. Index. 144pp. 8⅜ × 11¼. 26723-7 Pa. $8.95

GOTHIC AND OLD ENGLISH ALPHABETS: 100 Complete Fonts, Dan X. Solo. Add power, elegance to posters, signs, other graphics with 100 stunning copyright-free alphabets: Blackstone, Dolbey, Germania, 97 more—including many lower-case, numerals, punctuation marks. 104pp. 8⅛ × 11. 24695-7 Pa. $8.95

HOW TO DO BEADWORK, Mary White. Fundamental book on craft from simple projects to five-bead chains and woven works. 106 illustrations. 142pp. 5⅜ × 8.
20697-1 Pa. $4.95

THE BOOK OF WOOD CARVING, Charles Marshall Sayers. Finest book for beginners discusses fundamentals and offers 34 designs. "Absolutely first rate . . . well thought out and well executed."—E. J. Tangerman. 118pp. 7¾ × 10⅝.
23654-4 Pa. $5.95

ILLUSTRATED CATALOG OF CIVIL WAR MILITARY GOODS: Union Army Weapons, Insignia, Uniform Accessories, and Other Equipment, Schuyler, Hartley, and Graham. Rare, profusely illustrated 1846 catalog includes Union Army uniform and dress regulations, arms and ammunition, coats, insignia, flags, swords, rifles, etc. 226 illustrations. 160pp. 9 × 12. 24939-5 Pa. $10.95

WOMEN'S FASHIONS OF THE EARLY 1900s: An Unabridged Republication of "New York Fashions, 1909," National Cloak & Suit Co. Rare catalog of mail-order fashions documents women's and children's clothing styles shortly after the turn of the century. Captions offer full descriptions, prices. Invaluable resource for fashion, costume historians. Approximately 725 illustrations. 128pp. 8⅜ × 11¼.
27276-1 Pa. $11.95

THE 1912 AND 1915 GUSTAV STICKLEY FURNITURE CATALOGS, Gustav Stickley. With over 200 detailed illustrations and descriptions, these two catalogs are essential reading and reference materials and identification guides for Stickley furniture. Captions cite materials, dimensions and prices. 112pp. 6½ × 9¼.
26676-1 Pa. $9.95

EARLY AMERICAN LOCOMOTIVES, John H. White, Jr. Finest locomotive engravings from early 19th century: historical (1804–74), main-line (after 1870), special, foreign, etc. 147 plates. 142pp. 11⅜ × 8¼. 22772-3 Pa. $10.95

THE TALL SHIPS OF TODAY IN PHOTOGRAPHS, Frank O. Braynard. Lavishly illustrated tribute to nearly 100 majestic contemporary sailing vessels: Amerigo Vespucci, Clearwater, Constitution, Eagle, Mayflower, Sea Cloud, Victory, many more. Authoritative captions provide statistics, background on each ship. 190 black-and-white photographs and illustrations. Introduction. 128pp. 8⅞ × 11¾. 27163-3 Pa. $13.95

CATALOG OF DOVER BOOKS

EARLY NINETEENTH-CENTURY CRAFTS AND TRADES, Peter Stockham (ed.). Extremely rare 1807 volume describes to youngsters the crafts and trades of the day: brickmaker, weaver, dressmaker, bookbinder, ropemaker, saddler, many more. Quaint prose, charming illustrations for each craft. 20 black-and-white line illustrations. 192pp. 4⅝ × 6. 27293-1 Pa. $4.95

VICTORIAN FASHIONS AND COSTUMES FROM HARPER'S BAZAR, 1867–1898, Stella Blum (ed.). Day costumes, evening wear, sports clothes, shoes, hats, other accessories in over 1,000 detailed engravings. 320pp. 9⅜ × 12¼.
22990-4 Pa. $13.95

GUSTAV STICKLEY, THE CRAFTSMAN, Mary Ann Smith. Superb study surveys broad scope of Stickley's achievement, especially in architecture. Design philosophy, rise and fall of the Craftsman empire, descriptions and floor plans for many Craftsman houses, more. 86 black-and-white halftones. 31 line illustrations. Introduction. 208pp. 6½ × 9¼. 27210-9 Pa. $9.95

THE LONG ISLAND RAIL ROAD IN EARLY PHOTOGRAPHS, Ron Ziel. Over 220 rare photos, informative text document origin (1844) and development of rail service on Long Island. Vintage views of early trains, locomotives, stations, passengers, crews, much more. Captions. 8⅞ × 11¾. 26301-0 Pa. $13.95

THE BOOK OF OLD SHIPS: From Egyptian Galleys to Clipper Ships, Henry B. Culver. Superb, authoritative history of sailing vessels, with 80 magnificent line illustrations. Galley, bark, caravel, longship, whaler, many more. Detailed, informative text on each vessel by noted naval historian. Introduction. 256pp. 5⅜ × 8½. 27332-6 Pa. $6.95

TEN BOOKS ON ARCHITECTURE, Vitruvius. The most important book ever written on architecture. Early Roman aesthetics, technology, classical orders, site selection, all other aspects. Morgan translation. 331pp. 5⅜ × 8½. 20645-9 Pa. $8.95

THE HUMAN FIGURE IN MOTION, Eadweard Muybridge. More than 4,500 stopped-action photos, in action series, showing undraped men, women, children jumping, lying down, throwing, sitting, wrestling, carrying, etc. 390pp. 7⅞ × 10⅝.
20204-6 Clothbd. $24.95

TREES OF THE EASTERN AND CENTRAL UNITED STATES AND CANADA, William M. Harlow. Best one-volume guide to 140 trees. Full descriptions, woodlore, range, etc. Over 600 illustrations. Handy size. 288pp. 4½ × 6⅜.
20395-6 Pa. $5.95

SONGS OF WESTERN BIRDS, Dr. Donald J. Borror. Complete song and call repertoire of 60 western species, including flycatchers, juncoes, cactus wrens, many more—includes fully illustrated booklet. Cassette and manual 99913-0 $8.95

GROWING AND USING HERBS AND SPICES, Milo Miloradovich. Versatile handbook provides all the information needed for cultivation and use of all the herbs and spices available in North America. 4 illustrations. Index. Glossary. 236pp. 5⅜ × 8½. 25058-X Pa. $6.95

BIG BOOK OF MAZES AND LABYRINTHS, Walter Shepherd. 50 mazes and labyrinths in all—classical, solid, ripple, and more—in one great volume. Perfect inexpensive puzzler for clever youngsters. Full solutions. 112pp. 8⅛ × 11.
22951-3 Pa. $4.95

CATALOG OF DOVER BOOKS

PIANO TUNING, J. Cree Fischer. Clearest, best book for beginner, amateur. Simple repairs, raising dropped notes, tuning by easy method of flattened fifths. No previous skills needed. 4 illustrations. 201pp. 5⅜ × 8½. 23267-0 Pa. $5.95

A SOURCE BOOK IN THEATRICAL HISTORY, A. M. Nagler. Contemporary observers on acting, directing, make-up, costuming, stage props, machinery, scene design, from Ancient Greece to Chekhov. 611pp. 5⅜ × 8½. 20515-0 Pa. $11.95

THE COMPLETE NONSENSE OF EDWARD LEAR, Edward Lear. All nonsense limericks, zany alphabets, Owl and Pussycat, songs, nonsense botany, etc., illustrated by Lear. Total of 320pp. 5⅜ × 8½. (USO) 20167-8 Pa. $6.95

VICTORIAN PARLOUR POETRY: An Annotated Anthology, Michael R. Turner. 117 gems by Longfellow, Tennyson, Browning, many lesser-known poets. "The Village Blacksmith," "Curfew Must Not Ring Tonight," "Only a Baby Small," dozens more, often difficult to find elsewhere. Index of poets, titles, first lines. xxiii + 325pp. 5⅜ × 8¼. 27044-0 Pa. $8.95

DUBLINERS, James Joyce. Fifteen stories offer vivid, tightly focused observations of the lives of Dublin's poorer classes. At least one, "The Dead," is considered a masterpiece. Reprinted complete and unabridged from standard edition. 160pp. 5³⁄₁₆ × 8¼. 26870-5 Pa. $1.00

THE HAUNTED MONASTERY and THE CHINESE MAZE MURDERS, Robert van Gulik. Two full novels by van Gulik, set in 7th-century China, continue adventures of Judge Dee and his companions. An evil Taoist monastery, seemingly supernatural events; overgrown topiary maze hides strange crimes. 27 illustrations. 328pp. 5⅜ × 8½. 23502-5 Pa. $7.95

THE BOOK OF THE SACRED MAGIC OF ABRAMELIN THE MAGE, translated by S. MacGregor Mathers. Medieval manuscript of ceremonial magic. Basic document in Aleister Crowley, Golden Dawn groups. 268pp. 5⅜ × 8½.
23211-5 Pa. $8.95

NEW RUSSIAN-ENGLISH AND ENGLISH-RUSSIAN DICTIONARY, M. A. O'Brien. This is a remarkably handy Russian dictionary, containing a surprising amount of information, including over 70,000 entries. 366pp. 4½ × 6⅛.
20208-9 Pa. $9.95

HISTORIC HOMES OF THE AMERICAN PRESIDENTS, Second, Revised Edition, Irvin Haas. A traveler's guide to American Presidential homes, most open to the public, depicting and describing homes occupied by every American President from George Washington to George Bush. With visiting hours, admission charges, travel routes. 175 photographs. Index. 160pp. 8¼ × 11. 26751-2 Pa. $10.95

NEW YORK IN THE FORTIES, Andreas Feininger. 162 brilliant photographs by the well-known photographer, formerly with *Life* magazine. Commuters, shoppers, Times Square at night, much else from city at its peak. Captions by John von Hartz. 181pp. 9¼ × 10¾. 23585-8 Pa. $12.95

INDIAN SIGN LANGUAGE, William Tomkins. Over 525 signs developed by Sioux and other tribes. Written instructions and diagrams. Also 290 pictographs. 111pp. 6⅛ × 9¼. 22029-X Pa. $3.50

CATALOG OF DOVER BOOKS

ANATOMY: A Complete Guide for Artists, Joseph Sheppard. A master of figure drawing shows artists how to render human anatomy convincingly. Over 460 illustrations. 224pp. 8⅜ × 11¼. 27279-6 Pa. $10.95

MEDIEVAL CALLIGRAPHY: Its History and Technique, Marc Drogin. Spirited history, comprehensive instruction manual covers 13 styles (ca. 4th century thru 15th). Excellent photographs; directions for duplicating medieval techniques with modern tools. 224pp. 8⅜ × 11¼. 26142-5 Pa. $11.95

DRIED FLOWERS: How to Prepare Them, Sarah Whitlock and Martha Rankin. Complete instructions on how to use silica gel, meal and borax, perlite aggregate, sand and borax, glycerine and water to create attractive permanent flower arrangements. 12 illustrations. 32pp. 5⅜ × 8½. 21802-3 Pa. $1.00

EASY-TO-MAKE BIRD FEEDERS FOR WOODWORKERS, Scott D. Campbell. Detailed, simple-to-use guide for designing, constructing, caring for and using feeders. Text, illustrations for 12 classic and contemporary designs. 96pp. 5⅜ × 8½. 25847-5 Pa. $2.95

OLD-TIME CRAFTS AND TRADES, Peter Stockham. An 1807 book created to teach children about crafts and trades open to them as future careers. It describes in detailed, nontechnical terms 24 different occupations, among them coachmaker, gardener, hairdresser, lacemaker, shoemaker, wheelwright, copper-plate printer, milliner, trunkmaker, merchant and brewer. Finely detailed engravings illustrate each occupation. 192pp. 4⅝ × 6. 27398-9 Pa. $4.95

THE HISTORY OF UNDERCLOTHES, C. Willett Cunnington and Phyllis Cunnington. Fascinating, well-documented survey covering six centuries of English undergarments, enhanced with over 100 illustrations: 12th-century laced-up bodice, footed long drawers (1795), 19th-century bustles, 19th-century corsets for men, Victorian "bust improvers," much more. 272pp. 5⅝ × 8¼. 27124-2 Pa. $9.95

ARTS AND CRAFTS FURNITURE: The Complete Brooks Catalog of 1912, Brooks Manufacturing Co. Photos and detailed descriptions of more than 150 now very collectible furniture designs from the Arts and Crafts movement depict davenports, settees, buffets, desks, tables, chairs, bedsteads, dressers and more, all built of solid, quarter-sawed oak. Invaluable for students and enthusiasts of antiques, Americana and the decorative arts. 80pp. 6½ × 9¼. 27471-3 Pa. $7.95

HOW WE INVENTED THE AIRPLANE: An Illustrated History, Orville Wright. Fascinating firsthand account covers early experiments, construction of planes and motors, first flights, much more. Introduction and commentary by Fred C. Kelly. 76 photographs. 96pp. 8¼ × 11. 25662-6 Pa. $8.95

THE ARTS OF THE SAILOR: Knotting, Splicing and Ropework, Hervey Garrett Smith. Indispensable shipboard reference covers tools, basic knots and useful hitches; handsewing and canvas work, more. Over 100 illustrations. Delightful reading for sea lovers. 256pp. 5⅝ × 8½. 26440-8 Pa. $7.95

FRANK LLOYD WRIGHT'S FALLINGWATER: The House and Its History, Second, Revised Edition, Donald Hoffmann. A total revision—both in text and illustrations—of the standard document on Fallingwater, the boldest, most personal architectural statement of Wright's mature years, updated with valuable new material from the recently opened Frank Lloyd Wright Archives. "Fascinating"—*The New York Times*. 116 illustrations. 128pp. 9¼ × 10¾. 27430-6 Pa. $10.95

PHOTOGRAPHIC SKETCHBOOK OF THE CIVIL WAR, Alexander Gardner. 100 photos taken on field during the Civil War. Famous shots of Manassas, Harper's Ferry, Lincoln, Richmond, slave pens, etc. 244pp. 10⅝ × 8¼.
22731-6 Pa. $9.95

FIVE ACRES AND INDEPENDENCE, Maurice G. Kains. Great back-to-the-land classic explains basics of self-sufficient farming. The one book to get. 95 illustrations. 397pp. 5⅜ × 8½. 20974-1 Pa. $7.95

SONGS OF EASTERN BIRDS, Dr. Donald J. Borror. Songs and calls of 60 species most common to eastern U.S.: warblers, woodpeckers, flycatchers, thrushes, larks, many more in high-quality recording. Cassette and manual 99912-2 $8.95

A MODERN HERBAL, Margaret Grieve. Much the fullest, most exact, most useful compilation of herbal material. Gigantic alphabetical encyclopedia, from aconite to zedoary, gives botanical information, medical properties, folklore, economic uses, much else. Indispensable to serious reader. 161 illustrations. 888pp. 6½ × 9¼. 2-vol. set. (USO) Vol. I: 22798-7 Pa. $9.95
Vol. II: 22799-5 Pa. $9.95

HIDDEN TREASURE MAZE BOOK, Dave Phillips. Solve 34 challenging mazes accompanied by heroic tales of adventure. Evil dragons, people-eating plants, bloodthirsty giants, many more dangerous adversaries lurk at every twist and turn. 34 mazes, stories, solutions. 48pp. 8¼ × 11. 24566-7 Pa. $2.95

LETTERS OF W. A. MOZART, Wolfgang A. Mozart. Remarkable letters show bawdy wit, humor, imagination, musical insights, contemporary musical world; includes some letters from Leopold Mozart. 276pp. 5⅜ × 8½. 22859-2 Pa. $7.95

BASIC PRINCIPLES OF CLASSICAL BALLET, Agrippina Vaganova. Great Russian theoretician, teacher explains methods for teaching classical ballet. 118 illustrations. 175pp. 5⅜ × 8½. 22036-2 Pa. $4.95

THE JUMPING FROG, Mark Twain. Revenge edition. The original story of The Celebrated Jumping Frog of Calaveras County, a hapless French translation, and Twain's hilarious "retranslation" from the French. 12 illustrations. 66pp. 5⅜ × 8½.
22686-7 Pa. $3.95

BEST REMEMBERED POEMS, Martin Gardner (ed.). The 126 poems in this superb collection of 19th- and 20th-century British and American verse range from Shelley's "To a Skylark" to the impassioned "Renascence" of Edna St. Vincent Millay and to Edward Lear's whimsical "The Owl and the Pussycat." 224pp. 5⅜ × 8½.
27165-X Pa. $4.95

COMPLETE SONNETS, William Shakespeare. Over 150 exquisite poems deal with love, friendship, the tyranny of time, beauty's evanescence, death and other themes in language of remarkable power, precision and beauty. Glossary of archaic terms. 80pp. 5³⁄₁₆ × 8¼. 26686-9 Pa. $1.00

BODIES IN A BOOKSHOP, R. T. Campbell. Challenging mystery of blackmail and murder with ingenious plot and superbly drawn characters. In the best tradition of British suspense fiction. 192pp. 5⅜ × 8½. 24720-1 Pa. $5.95

THE WIT AND HUMOR OF OSCAR WILDE, Alvin Redman (ed.). More than 1,000 ripostes, paradoxes, wisecracks: Work is the curse of the drinking classes; I can resist everything except temptation; etc. 258pp. 5⅜ × 8½. **20602-5 Pa. $5.95**

SHAKESPEARE LEXICON AND QUOTATION DICTIONARY, Alexander Schmidt. Full definitions, locations, shades of meaning in every word in plays and poems. More than 50,000 exact quotations. 1,485pp. 6½ × 9¼. 2-vol. set.
Vol. I: **22726-X Pa. $16.95**
Vol. 2: **22727-8 Pa. $15.95**

SELECTED POEMS, Emily Dickinson. Over 100 best-known, best-loved poems by one of America's foremost poets, reprinted from authoritative early editions. No comparable edition at this price. Index of first lines. 64pp. 5 3/16 × 8¼. **26466-1 Pa. $1.00**

CELEBRATED CASES OF JUDGE DEE (DEE GOONG AN), translated by Robert van Gulik. Authentic 18th-century Chinese detective novel; Dee and associates solve three interlocked cases. Led to van Gulik's own stories with same characters. Extensive introduction. 9 illustrations. 237pp. 5⅜ × 8½. **23337-5 Pa. $6.95**

THE MALLEUS MALEFICARUM OF KRAMER AND SPRENGER, translated by Montague Summers. Full text of most important witchhunter's "bible," used by both Catholics and Protestants. 278pp. 6⅝ × 10. **22802-9 Pa. $11.95**

SPANISH STORIES/CUENTOS ESPAÑOLES: A Dual-Language Book, Angel Flores (ed.). Unique format offers 13 great stories in Spanish by Cervantes, Borges, others. Faithful English translations on facing pages. 352pp. 5⅜ × 8½. **25399-6 Pa. $8.95**

THE CHICAGO WORLD'S FAIR OF 1893: A Photographic Record, Stanley Appelbaum (ed.). 128 rare photos show 200 buildings, Beaux-Arts architecture, Midway, original Ferris Wheel, Edison's kinetoscope, more. Architectural emphasis; full text. 116pp. 8¼ × 11. **23990-X Pa. $9.95**

OLD QUEENS, N.Y., IN EARLY PHOTOGRAPHS, Vincent F. Seyfried and William Asadorian. Over 160 rare photographs of Maspeth, Jamaica, Jackson Heights, and other areas. Vintage views of DeWitt Clinton mansion, 1939 World's Fair and more. Captions. 192pp. 8⅞ × 11. **26358-4 Pa. $12.95**

CAPTURED BY THE INDIANS: 15 Firsthand Accounts, 1750–1870, Frederick Drimmer. Astounding true historical accounts of grisly torture, bloody conflicts, relentless pursuits, miraculous escapes and more, by people who lived to tell the tale. 384pp. 5⅜ × 8½. **24901-8 Pa. $8.95**

THE WORLD'S GREAT SPEECHES, Lewis Copeland and Lawrence W. Lamm (eds.). Vast collection of 278 speeches of Greeks to 1970. Powerful and effective models; unique look at history. 842pp. 5⅜ × 8½. **20468-5 Pa. $14.95**

THE BOOK OF THE SWORD, Sir Richard F. Burton. Great Victorian scholar/adventurer's eloquent, erudite history of the "queen of weapons"—from prehistory to early Roman Empire. Evolution and development of early swords, variations (sabre, broadsword, cutlass, scimitar, etc.), much more. 336pp. 6⅛ × 9¼. **25434-8 Pa. $8.95**

AUTOBIOGRAPHY: The Story of My Experiments with Truth, Mohandas K. Gandhi. Boyhood, legal studies, purification, the growth of the Satyagraha (nonviolent protest) movement. Critical, inspiring work of the man responsible for the freedom of India. 480pp. 5⅜ × 8½. (USO) 24593-4 Pa. $8.95

CELTIC MYTHS AND LEGENDS, T. W. Rolleston. Masterful retelling of Irish and Welsh stories and tales. Cuchulain, King Arthur, Deirdre, the Grail, many more. First paperback edition. 58 full-page illustrations. 512pp. 5⅜ × 8½. 26507-2 Pa. $9.95

THE PRINCIPLES OF PSYCHOLOGY, William James. Famous long course complete, unabridged. Stream of thought, time perception, memory, experimental methods; great work decades ahead of its time. 94 figures. 1,391pp. 5⅜ × 8½. 2-vol. set.
Vol. I: 20381-6 Pa. $12.95
Vol. II: 20382-4 Pa. $12.95

THE WORLD AS WILL AND REPRESENTATION, Arthur Schopenhauer. Definitive English translation of Schopenhauer's life work, correcting more than 1,000 errors, omissions in earlier translations. Translated by E. F. J. Payne. Total of 1,269pp. 5⅜ × 8½. 2-vol. set.
Vol. 1: 21761-2 Pa. $11.95
Vol. 2: 21762-0 Pa. $11.95

MAGIC AND MYSTERY IN TIBET, Madame Alexandra David-Neel. Experiences among lamas, magicians, sages, sorcerers, Bonpa wizards. A true psychic discovery. 32 illustrations. 321pp. 5⅜ × 8½. (USO) 22682-4 Pa. $8.95

THE EGYPTIAN BOOK OF THE DEAD, E. A. Wallis Budge. Complete reproduction of Ani's papyrus, finest ever found. Full hieroglyphic text, interlinear transliteration, word-for-word translation, smooth translation. 533pp. 6½ × 9¼. 21866-X Pa. $9.95

MATHEMATICS FOR THE NONMATHEMATICIAN, Morris Kline. Detailed, college-level treatment of mathematics in cultural and historical context, with numerous exercises. Recommended Reading Lists. Tables. Numerous figures. 641pp. 5⅜ × 8½. 24823-2 Pa. $11.95

THEORY OF WING SECTIONS: Including a Summary of Airfoil Data, Ira H. Abbott and A. E. von Doenhoff. Concise compilation of subsonic aerodynamic characteristics of NACA wing sections, plus description of theory. 350pp. of tables. 693pp. 5⅜ × 8½. 60586-8 Pa. $14.95

THE RIME OF THE ANCIENT MARINER, Gustave Doré, S. T. Coleridge. Doré's finest work; 34 plates capture moods, subtleties of poem. Flawless full-size reproductions printed on facing pages with authoritative text of poem. "Beautiful. Simply beautiful."—*Publisher's Weekly*. 77pp. 9¼ × 12. 22305-1 Pa. $6.95

NORTH AMERICAN INDIAN DESIGNS FOR ARTISTS AND CRAFTSPEOPLE, Eva Wilson. Over 360 authentic copyright-free designs adapted from Navajo blankets, Hopi pottery, Sioux buffalo hides, more. Geometrics, symbolic figures, plant and animal motifs, etc. 128pp. 8⅜ × 11. (EUK) 25341-4 Pa. $7.95

SCULPTURE: Principles and Practice, Louis Slobodkin. Step-by-step approach to clay, plaster, metals, stone; classical and modern. 253 drawings, photos. 255pp. 8⅛ × 11. 22960-2 Pa. $10.95

CATALOG OF DOVER BOOKS

THE INFLUENCE OF SEA POWER UPON HISTORY, 1660-1783, A. T. Mahan. Influential classic of naval history and tactics still used as text in war colleges. First paperback edition. 4 maps. 24 battle plans. 640pp. 5⅜ × 8½.
25509-3 Pa. $12.95

THE STORY OF THE TITANIC AS TOLD BY ITS SURVIVORS, Jack Winocour (ed.). What it was really like. Panic, despair, shocking inefficiency, and a little heroism. More thrilling than any fictional account. 26 illustrations. 320pp. 5⅜ × 8½.
20610-6 Pa. $8.95

FAIRY AND FOLK TALES OF THE IRISH PEASANTRY, William Butler Yeats (ed.). Treasury of 64 tales from the twilight world of Celtic myth and legend: "The Soul Cages," "The Kildare Pooka," "King O'Toole and his Goose," many more. Introduction and Notes by W. B. Yeats. 352pp. 5⅜ × 8½. 26941-8 Pa. $8.95

BUDDHIST MAHAYANA TEXTS, E. B. Cowell and Others (eds.). Superb, accurate translations of basic documents in Mahayana Buddhism, highly important in history of religions. The Buddha-karita of Asvaghosha, Larger Sukhavativyuha, more. 448pp. 5⅜ × 8½. ,
25552-2 Pa. $9.95

ONE TWO THREE . . . INFINITY: Facts and Speculations of Science, George Gamow. Great physicist's fascinating, readable overview of contemporary science: number theory, relativity, fourth dimension, entropy, genes, atomic structure, much more. 128 illustrations. Index. 352pp. 5⅜ × 8½. 25664-2 Pa. $8.95

ENGINEERING IN HISTORY, Richard Shelton Kirby, et al. Broad, nontechnical survey of history's major technological advances: birth of Greek science, industrial revolution, electricity and applied science, 20th-century automation, much more. 181 illustrations. ". . . excellent . . ."—Isis. Bibliography. vii + 530pp. 5⅜ × 8¼.
26412-2 Pa. $14.95

Prices subject to change without notice.

Available at your book dealer or write for free catalog to Dept. GI, Dover Publications, Inc., 31 East 2nd St., Mineola, N.Y. 11501. Dover publishes more than 500 books each year on science, elementary and advanced mathematics, biology, music, art, literary history, social sciences and other areas.